Reflections on World Affairs

Peace and Politics

Essential Reading Series on Eastern Culture, Civilization and History — Volume 1

Reflections on World Affairs

Peace and Politics

Ahmed H. Zewail

Nobel Laureate

California Institute of Technology, USA

Published by

Imperial College Press
57 Shelton Street
Covent Garden
London WC2H 9HE

Distributed by

World Scientific Publishing Co. Pte. Ltd.

5 Toh Tuck Link, Singapore 596224

USA office: 27 Warren Street, Suite 401-402, Hackensack, NJ 07601

UK office: 57 Shelton Street, Covent Garden, London WC2H 9HE

Library of Congress Cataloging-in-Publication Data
Zewail, Ahmed H.
   Reflections on world affairs : peace and politics / Ahmed H. Zewail (Nobel laureate, California Institute of Technology).
      pages cm -- (Essential reading series on Eastern culture, civilization and history ; vol. 1)
   Includes bibliographical references and index.
   ISBN 978-1-78326-725-5 (hardcover : alkaline paper) -- ISBN 1-78326-725-9 (hardcover : alkaline paper) --
   ISBN 978-1-78326-726-2 (paperback : alkaline paper) -- ISBN 1-78326-726-7 (paperback : alkaline paper)
   1. World politics--21st century. 2. International relations. 3. World politics--Philosophy. 4. International relations--Philosophy. 5. Peace--Political aspects. 6. Education and state. 7. Science and state. 8. Technology and state. 9. Egypt--Foreign relations--21st century. 10. United States--Foreign relations--21st century. I. Title.
   D863.Z49 2015
   909.83'1--dc23

2014049713

British Library Cataloguing-in-Publication Data
A catalogue record for this book is available from the British Library.

Cover concept and design by Drs. Zewail and Shorokhov.

Printed in Singapore

To Egypt

where my roots were planted

To America

where the opportunity began

To the World

where commitment to world-peace bloomed

Preface

Today, our world seems to be falling apart. Russian troops are near the border with Ukraine. Tensions are building in the South China Sea. We are witnessing a bloody civil war in Syria and Iraq. The Palestinian-Israeli conflict has escalated, with thousands of innocent people injured or killed, and the prospect of permanent peace is not on the horizon. The so-called Arab Spring that many hoped would sweep the Middle East is now yielding to the "Arab Awakening" that proponents claim, through "creative chaos," will result in a democratic Arab World. More alarming is the spread of militant groups such as Al-Qaeda, Jabhat al-Nusra, and the Islamic State of Iraq and Syria (ISIS) that may acquire nuclear capabilities and draw the superpowers into major wars. The period before WWI, which broke out in 1914, was known as *La Belle Époque*, or *The Beautiful Age*. A century later, these flashpoints of conflict are now defining an age of a different kind.

Is peace beyond our reach, and what does the future hold? Can we provide basic education to all children? Why the decline in the Arab and Muslim world after so much past achievement? Is Islam the problem? In the land of opportunity—the United States of America—can the Republic maintain world leadership? Can the U.S. sustain its leadership in innovation and prosperity, given the evolution of its culture and politics and the rise of other superpowers? And, in this century, how does knowledge acquisition through education and scientific research determine the *wealth of nations*?

These are big questions for which no one has the full answers, but I have endeavored to address the pertinent issues. Through my writings in op-eds and essays, and in public lectures, I have presented a dissection of the problems involved in an attempt to reduce them to their basic elements. These elements, I have found, are rather simple despite the complexity of the world we inhabit. From my experience as the first United States Science Envoy to the Middle East, it is clear that most problems in the region, as elsewhere in the world, arise from "not knowing" and "not having." It follows that education is critical, not only for enlightenment, or "knowing," but also for boosting productivity and enhancing "having."

My concern with these issues has a focus on global science education and science in diplomacy. I believe that the use of the soft power of education, diplomacy, and economic development is far more effective, and much less expensive, than the hegemony of military aggression or politicized foreign aid.

Over the years, I have devoted a significant amount of effort to helping the world and culture of my youth (the "East"), where I received my early and extraordinary education, and the world of my professional development (the "West"), where I am provided with optimum opportunities. Remarkably, I did not experience a conflict of cultures, or the so-called "conflict of civilizations" between the East and the West, a concept popularized by some scholars and advanced as inevitable, especially since the tragedy of September 11, 2001.

From my unique position straddling East and West cultures and values, and as the only living recipient of the Nobel Prize in Science or Medicine in the Arab and Muslim world, it was possible to build bridges of modern knowledge and to finally establish a major center of excellence in Cairo, named by Egypt's government, the Zewail City of Science and Technology. To establish the City as an independent entity took 15 years of hard work and outreach to 4 presidents, 8 prime ministers, and more than 20 ministers! Equally important is my endeavor of informing the public for nearly two decades that has resulted in the inspiration of a large segment of the Arab population by promoting the value of useful knowledge in development and progress, accomplished in part through TV addresses to millions of viewers and by means of publications, such as *Asr Al Alm*, or *The Epoch of Science* book, now in its 17th edition.

This volume, *The Collected Work,* is an assemblage of my reflections and thoughts on the issues pertinent to global affairs. Many have helped me formulate my thoughts, from Mohamed, a driver in Cairo, to Esraa, a Tahrir Square youth revolutionary, and from my audience in public discourses all over the world to the discussions I have had with government leaders in the Muslim World, including Prime Minister Dr. Mahathir Mohamad of Malaysia, and Vice President and Prime Minister of the United Arab Emirates and Ruler of Dubai, Sheikh Mohammed bin Rashid Al Maktoum. Some of the predictions I made are still robust, and some have evolved with time, while some others did not transpire, either because of "politics as usual" or the emergence of sudden, cataclysmic events in the Middle East.

To those who made possible the production of this volume in its present form, I am grateful. My colleague at Caltech, Dr. Dmitry Shorokhov, has had a real impact, not only because of his technical knowledge, but also for the stimulating weekend discussions on the unhappy aftermath of revolutions, comparing and contrasting the Arab Spring with Bolshevism in Russia. Without the exceptional efficiency of the publisher, Imperial College Press, and its leader Dr. K. K. Phua, this Work would not have been realized and printed on time: he surely knows how to lure me into writing or composing a new volume! To Ms. Sook Cheng Lim goes my gratitude for her care and promptness.

I have been aided in my offices in Pasadena and Cairo by a "trio" of assistants at Caltech and in Zewail City who, despite the pressure of continuous and "urgent" administrative issues, have done everything possible to shield me so that I could have hours of peace for my science and public affairs work. To Ms. De Ann Lewis, Ms. Magnolia (Maggie) Sabanpan, and Ms. Ragia Mansour, I am indebted for their sincere efforts. De Ann's valuable input to some pieces in this volume is greatly appreciated.

Last, but not least, is the importance of my "sounding board" at home. My wife, Dema, is a critical reviewer; indeed, she has made a significant contribution, in particular to some op-eds on the Middle East and global affairs. I do hope that she and my children have enjoyed the experience as much as I have, and that in this *Collected Work* they may find the beauty of bearing the two cultures they have inherited.

Ahmed Zewail
August 18, 2014

Contents[*]

[*] *These are the Contents at a glance; more details can be found at the end of the book.*

Chapter 1

Egypt and the Muslim World

Opinions-Editorials
(Op-Eds)*

* References to these Articles are given in the Index at the end of the book.

THE HUFFINGTON POST

TOP NEWS AND OPINION

Published: 5 November 2014

Why It Would Be a Big Mistake for the U.S. to Cut Aid to Egypt

Today, the U.S. needs Egypt's partnership more than ever.

Op-Ed by Ahmed H. Zewail

CAIRO — Echoing calls by some in the U.S. Congress, the New York Times editorial board recently published a piece arguing that American aid to Egypt should be cut as a way of punishing President Abdel Fattah el-Sisi and his administration. As an Egyptian-American and someone well immersed in both cultures, I believe that such action would prompt a tipping point in the U.S.-Egypt relationship and have serious consequences for the future of the Middle East.

Some weeks ago I met with President Sisi and former President Adly Mansour, along with key officials including the prime minister, Mr. Ibrahim Mehleb. I also had a chance to speak to a gathering of university students and to meet with leading groups of independent and government-run media, including a TV interview that was watched by millions of Egyptians. Throughout the two weeks of intensive discussions, I came to understand the basis for the potential change in the relationship, and why the majority of people support Sisi.

In the period following the Egyptian Revolution of 1952, the time of my schooling in Egypt, U.S. Secretary of State John Foster Dulles sought to punish President Gamal Abdel Nasser, and the U.S. decided not to support the building of the Aswan Dam and create a source of hydroelectric power considered pivotal to Egypt's industrialization. The result? Egypt's political compass swung from West to East, and the Soviet Union had a major influence in the Middle East for decades.

President Anwar Sadat reversed that direction in 1973, and for 40 years, the Middle East has witnessed peace between Israel and Egypt. The current political temperature in Cairo is similar to that of Nasser's era, and in fact an analogy is often made between Nasser's and Sisi's popularity.

Today, the U.S. needs Egypt's partnership more than ever. In addition to the peace treaty between Egypt and Israel, which is crucial to U.S. interests both domestically and in the Middle East, the U.S. has had and will continue to need Egypt's collaboration in the war on terrorism. The U.S. has full access to the Suez Canal, and the military joint exercises already in existence are necessary for such wars and for the free flow of oil. Last month, northern Sinai was struck by terrorists groups, killing more than 30 Army personnel and wounding tens of innocent Egyptians. The Islamic State to the country's east must be stopped from getting into the Sinai and the oil fields in Iraq and Saudi Arabia.

Knowing these facts, Egyptians reject the political manipulation that often accompanies the $1.3 billion aid. Moreover, the aid that now comes to Egypt from the Gulf States amounts to more than 10 times that from the U.S. In an interdependent world, the market for and diversity in military equipment could change Egypt's special relationship with the U.S.

With the rise of fanaticism in the region and the experience with the Muslim Brotherhood, the Egyptian population at large became fearful of the Brotherhood's return into governance, and many Egyptians see Sisi as a savior in this regard. When Mohamed Morsi was elected president of Egypt in 2012, many in the country, including me, were hopeful that he would become a democratic president for all Egyptians — not only for the Muslim Brotherhood. Unfortunately, his presidency quickly became a proxy for the Muslim Brotherhood, and under his leadership the country was driven to the edge of civil war. Millions took to the streets on June 30, 2013, and a group of civil liberals, religious, and military leaders led by Mr. Sisi decreed, after the isolation of Dr. Morsi, a new roadmap of change involving the election of a new president and parliament and the writing of a new constitution.

"Mr. El Sisi did not want the job," Adli Mansour, the chief justice of Egypt's Supreme Court and the president who succeeded Morsi, told me. "At the end, he decided to run because the people wanted him, and I told him it was his destiny," he added.

If the election was rigged, as some politicians and editorials have asserted, why would Egyptians continue to support him even after the election?

Shortly after Sisi was elected, his administration announced cuts of "subsidies" on natural gas and energy consumption and lowered those for bread and other goods. Such action was taboo during the Mubarak and Sadat presidencies for over half a century, but Sisi was able to convince Egyptians he was taking necessary action.

In another post-election call to Egyptians, he proclaimed the inauguration of a national project — the New Suez Canal — a waterway parallel to the old one dug in 1869, and called on Egyptians to invest in the project. In eight days, the Central Bank of Egypt raised EGP 61 billion (nearly $8.5 billion) by selling investment certificates. I visited one bank during those eight days and the lines circled several blocks. The bank stayed open late due to the unexpected huge volume of transactions.

It is true that Egypt's attempt at democracy after the 2011 revolution encountered many obstacles in governance and infrastructure. Many in the media were overzealous to then Field Marshal Sisi when he ran for the presidency after Mansour. And, since the 2013 revolution, there remain issues to address, among them the rule of law for NGOs, the cases of political prisoners awaiting trials, and the integration of the Muslim Brotherhood population into the political fabric of Egypt.

At this pivotal time, the U.S. should assist Egypt through direct dialogue and partnership. The leverage is America's soft power, access to the American market, a free trade agreement and the aid to build new educational and democratic institutions. The so-called Arab Spring has proved that the fall of a Mubarak-like presidency does not mean the immediate rise of democracy. In spite of this, I am confident that Egypt will not return to an authoritarian governing system again, and that with some time, it will achieve its democratic goals.

Egypt is facing monumental problems. Besides the issues of security to its east (the Islamic State), to its west (Libya), and in the south (Yemen), there are internal issues — economic and unemployment factors — of grave concern. But despite this, Sisi has managed to get the majority of Egyptians behind him, taken serious steps toward reforming the ailing economy, and given hope to the country by initiating major national projects, such as the New Suez Canal and the new City of Science and Technology. He is the first president to form a Council of Advisors of scientists and engineers to aid him in solving major national problems. As the Economist put it in a piece about Sisi's first 100 days, the president, "has brought economic and diplomatic advances as well as hope to Egyptians wearied by years of political turmoil."

Threatening Egypt with aid cuts is not in the best interest of the U.S.-Egypt relationship. The issue is no longer Sisi alone. Rather, it is "We the People" who are also deciding on future relationships, not only with the U.S., but also with Israel.

By Ahmed H. Zewail, the 1999 Nobel Laureate in Chemistry. Zewail is a professor at Caltech and Chairman of the Board of Egypt's City of Science and Technology.

Los Angeles Times

Published: 3 November 2014

Don't Cut Aid to Egypt: The Hopeful Case for Supporting Egyptian President Sisi

Today, the U.S. needs Egypt's partnership more than ever.

Op-Ed by Ahmed H. Zewail

Some members of Congress have criticized Egyptian President Abdel Fattah Sisi lately and called for a reduction in or elimination of U.S. military aid as a way of punishing his administration. After meeting with Sisi in Cairo recently and talking to a wide range of citizens there, I have come to understand why most Egyptians now support him. And I believe that cutting foreign aid to Egypt at this point would harm the U.S.-Egypt relationship and have serious consequences for the Middle East.

Abdel Fattah Sisi, center, has taken serious steps toward reforming the ailing Egyptian economy. (Pablo Martinez Monsivais / AP)

History illustrates the danger. In 1955, in the wake of the Egyptian Revolution of 1952, the United States agreed to provide funding to help build the Aswan Dam and create a source of hydroelectric power considered pivotal to Egypt's industrialization. Then, just months later, Secretary of State John Foster Dulles became convinced that Egypt's president, Gamal Abdel Nasser, wasn't trustworthy, and he withdrew the U.S. offer of funding. The result: Egypt's political compass swung from West to East, and the Soviet Union quickly stepped in to fill the void.

It wasn't until 1973 that the direction was reversed by President Anwar Sadat. In the 40 years since then, the U.S.-Egypt relationship has been extremely important, and for 40 years the Middle East has witnessed peace between Israel and Egypt.

Today, the U.S. needs Egypt's partnership more than ever. In addition to the peace treaty between Egypt and Israel, which is crucial to U.S. interests both domestically and in the Middle East, the U.S. has had and will continue to need Egypt's collaboration in the war on terrorism. Just last month, northern Sinai was struck by terrorists, who killed more than 30 Egyptian army personnel and wounded a number of civilians.

The partnership between the United States and Egypt is crucial to both countries, and it can't be predicated on political manipulation and threats of withholding aid. Moreover, the United States must be aware that it is no longer the primary provider of foreign aid to Egypt. Today, the Gulf States contribute more than 10 times what the U.S. does.

When Mohamed Morsi was elected president of Egypt in 2012, many in the country, including me, were hopeful that he would become a democratic president for all Egyptians. Unfortunately, his presidency quickly became a proxy for the Muslim Brotherhood, and under his leadership the country was driven to the edge of civil war. Millions took to the streets on June 30, 2013, to demand change and greater stability for Egypt.

President Sisi did not initially intend to run for the office in which he now serves, but he was urged to, I was told, by the chief justice of Egypt's Supreme Court and others. If the election that put him into office was rigged, as some politicians and editorials have claimed, why would Egyptians continue to support him after the election?

It is certainly not because he has taken the path of political expediency. Shortly after Sisi was elected, his administration announced cuts of "subsidies" on natural gas and energy consumption and lowered those for bread and other goods. This was an important step for economic stability in Egypt, but was considered politically impossible for more than half a century during the presidencies of Sadat and Hosni Mubarak. Sisi was able to convince Egyptians he was taking necessary action.

In another post-election call to Egyptians, he proclaimed the inauguration of a national project — the New Suez Canal — a waterway parallel to the one dug in 1869, and he called on Egyptians to invest in the project. In eight days, the Central Bank of Egypt raised nearly $8.5 billion by selling investment certificates. I visited one bank during those eight days, and the line circled several blocks. Banks had to stay open late to handle the unexpectedly huge volume of transactions.

It is true that Egypt's attempt at democracy after the 2011 revolution encountered many obstacles. And there remain issues to address, among them establishing fair laws governing NGOs, enforcing the rule of law for political prisoners awaiting trials, and the integration of Muslim Brotherhood members into the political fabric of Egypt.

These issues make it all the more crucial, however, for the U.S. to continue to engage Egypt through direct dialogue and partnership. America should not hesitate to wield its considerable soft power — providing access to American markets, initiating trade agreements, providing aid for building new educational and democratic institutions. The so-called Arab Spring has proved that the fall of a Mubarak-like presidency does not mean the immediate rise of democracy. That will take time and nurturing and encouragement.

Egypt is facing monumental problems. Besides internal issues, including a troubled economy and high unemployment, it has security problems to its east with Islamic State, to its west with Libya and in the south toward Yemen. But despite these issues, Sisi has managed to get the majority of Egyptians behind him, taken serious steps toward reforming the ailing economy, and given hope to the country by initiating major national projects, including the City of Science and Technology, which I have been actively involved in promoting for many years. As the Economist put it in a piece about Sisi's first 100 days, the president "has brought economic and diplomatic advances as well as hope to Egyptians wearied by years of political turmoil."

The U.S. needs to feed that hope, and cutting aid to Egypt won't accomplish that.

Ahmed H. Zewail, the 1999 Nobel laureate in chemistry, is a professor at Caltech and chairman of the board of trustees of Egypt's City of Science and Technology.

© 2014 The Los Angeles Times

The New York Times

Published: 13 October 2013

The Revolution Egypt Needs

Op-Ed by Ahmed H. Zewail

Pasadena, California — When I was a boy in Desuq, Egypt, a city on the Rosetta branch of the Nile, about 50 miles east of Alexandria, my family lived steps away from the local landmark, a mosque named for a 13th-century Sufi sheik. Five times a day, we would hear the call to prayer. Our imam encouraged us to study, telling my friends and me, again and again, of the message revealed by the Prophet Muhammad: "*iqra*" — read! Education was in the fabric of our culture and religion.

I left Egypt in 1969 for graduate school at the University of Pennsylvania. I have been on the faculty at Caltech for 37 years and carried dual citizenship for 31. But my commitment to the country of my birth never wavered. Political tumult — two uprisings, and the overthrow of two regimes, in the space of two years — has left Egypt in deep political uncertainty. But what's been lost in the deadly machinations among both the secular liberals and political Islamists is what touched off the revolution: the aspirations of Egypt's youth.

Like many Arab societies, Egypt is young. The activists who filled Tahrir Square in 2011 demanded liberty and social justice, valid ends in themselves, but their ultimate goal, I believe, was social and economic change — educational opportunities, leading to sound jobs and a decent life — necessary to flourish in the modern world. As the first Egyptian, and Arab, to be awarded a Nobel Prize in science, and a former special envoy sent by the Obama administration to promote science in the Middle East, this is my foremost concern.

Westerners often forget Egypt's long history of educational accomplishment. Al Azhar University, a center of Islamic learning, predates Oxford and Cambridge by centuries. Cairo University, founded in 1908, has been a center of enlightenment for the whole Arab world. Intellectuals pioneered Egypt's first democratic elections in the 1920s through the 1950s, under the monarchy that succeeded British rule. This period of modernization, into which I was born, included the establishment of scientific institutions and the emergence of modern industries like banking, news media, textiles and motion pictures.

I grew up during the time of Gamal Abdel Nasser, who participated in the 1952 revolt that overthrew the monarchy and led the country until his death in 1970. His was a state deficient in democracy, but not in optimism. Science, engineering and technology were among the top-ranked disciplines in the country's universities, which attracted the best students and scholars from the Arab world. Huge infrastructure projects, like the high dam at Aswan and the nuclear reactor in Inchass, required skilled engineers, which Egypt was able to provide. As an instructor at Alexandria University, I did research that was published in international journals. Although I left to pursue a doctorate in the United States, it was not for want of a good life.

But in the past 30 years, roughly since the assassination of Anwar el-Sadat, Nasser's successor, the country deteriorated. During the rule of President Hosni Mubarak, attention to schools and infrastructure gave way to a focus on media and security, mega resorts and vanity projects, even as a growing population produced intense — and unfulfilled — demands for education.

It gives me no pride to note that in science and technology, Egypt, and the entire Arab world, have made insignificant contributions. A part of the world that pioneered science and mathematics during Europe's dark ages is now lost in a dark age of illiteracy and knowledge deficiency. With the exception of Israel, the region's scientific output is modest at best. Turkey and Iran have made strides in technology; Egypt under Mubarak, in contrast, depended on revenues from the Suez Canal, tourism, gas and oil, with little contribution from high-tech industries.

After Mubarak was overthrown, Essam Sharaf, who was prime minister for less than a year, called on me to establish what the government named the Zewail City of Science and Technology, an educational and research project I had proposed to Mubarak and a number of prime ministers for nearly 15 years, without success. With immense public support, we raised money to create the project on more than 100 acres on the outskirts of Cairo. The leaders governing Egypt since the latest uprising, in June, have continued to support it. Essam Heggy, a planetary scientist at the NASA Jet Propulsion Laboratory and an adviser to the interim president, Adly Mansour, recently said that "education and science must be our national priority."

Research in biomedicine, solar energy, nanotechnology and other fields is under way. Last summer, some 6,000 applicants applied for spots at the university. I continue to support the project and to lead its board of trustees, which includes six Nobel laureates, but violence must end for the project to succeed. And high unemployment among young people, who represent nearly one-third of Egypt's population of 90 million, all but guarantees instability.

Egypt is strategically vital for the United States, because of the Suez Canal, its peace treaty with Israel and its cooperation with the American military and intelligence agencies. But most of the discussion about aid has focused on political leverage. America should instead think of aid in new, apolitical ways. The U.S. gives about $1.5 billion a year to Egypt and $3 billion to Israel; the former goes mainly for military equipment, while the latter is more of a partnership that includes not just military but also scientific and industrial cooperation.

I call on Egypt's leaders, of whatever religious or political persuasion, to insulate education and science from their feuds. I also call on great powers like the U.S. to support the development of human capital. The aid America gave Japan, South Korea and Taiwan after World War II, for example, enabled them to become economically vital.

I remain optimistic about Egypt, whose people will no longer settle for the status quo of the past half-century. The question, one I cannot answer as a scientist, is what will replace it, and how long it will take. Egyptians are known for their patience, which derives, perhaps, from the eternity of the Nile. But their patience has run thin, and their aspirations are unmet. Any group hoping to authentically represent the hopes of the Egyptian people must make educational attainment and economic growth its priority.

Ahmed Zewail, a professor of chemistry and physics at the California Institute of Technology, was awarded the Nobel Prize in chemistry in 1999.

The CHRISTIAN SCIENCE
MONITOR

Published: 11 July 2013

Healing Egypt: Three Steps to Unify a Divided Nation

The uprising of millions of Egyptians since June 30 has led to sharp polarization. Growing up in Egypt, I never saw the country as divided as it is today. Efforts to rebuild the nation must focus on justice, reconciliation, and inclusiveness.

Op-Ed by Ahmed H. Zewail

The uprising of millions of Egyptians since June 30 has led to sharp polarization. Some consider the removal of Mohamed Morsi a coup by the army against an elected president. Others treat it as the second revolution, or the continuation of the January 25, 2011, revolution. The media, especially in the West, is mainly concerned with the definition of a coup and whether the military should be punished by stopping US aid to Egypt.

The picture is not this simple, and the current situation is more than a coup definition; it is the healing of a country that has enormous potential and strategic position in the already troubled Middle East.

Supporters of Egypt's ousted president, Mohamed Morsi, protest before breaking their fast during the Islamic month of Ramadan, in Nasr City, Cairo, July 10. Op-ed contributor Ahmed Zewail writes: 'No civil, liberal government can succeed, even after new elections, if the Islamists are forced to work underground as a foe and the country remains divided.'

(Manu Brabo)

The real question is: What can be done for Egypt in its democratic transition, with Egyptians being strongly polarized? The proposed immediate action plan presented here can change the current situation and make the country move forward.

Growing up in Egypt, I never saw the country as divided as it is today. We now have two main political groupings: the Islamist parties and the civil, or liberal, political parties. What is also new is the youth movement – more powerful than present liberal parties – that uses the latest in technological tools to lead these street uprisings (millionea) because they want to live in a developed and prosperous Egypt.

On the one hand, the liberals in the country believe that the Muslim Brotherhood failed the democratic process because, although Mr. Morsi won the popular vote, he did not succeed in uniting the nation and serving as president of all Egyptians.

Morsi's appointment of Muslim Brotherhood members in leading positions of the county (so-called "*Akhwanet Egypt*"), his unexpected constitutional decrees, and his insistence on keeping a government seen by many as incompetent all were issues that led the youthful "rebel movement" to collect more than 20 million signatures calling for early presidential elections to remove him.

On the other hand, supporters of Morsi believe that he came to power democratically as the first elected civilian president in Egypt's history. As such, he should only have been able to be removed after completing his term in office. Only such a course, in their view, would protect the constitution (*hematite el-sharia*) that was passed by a national referendum with two-thirds majority. The Brotherhood's Freedom and Justice Party was a majority in the elected parliament.

The Egyptian army had one of two options: either defend Morsi's claim to power, leaving millions on the streets and the country as a whole sinking economically, and having national security being threatened with chaos; or interfere and put the country on a new course without being directly involved in governance.

So far, the latter is what the army has chosen to do. Unlike the ruling Supreme Council of the Armed Forces in the January 25 revolution, First General Abdel-Fattah El-Sisi remains without a political title. The chief justice of Egypt has been sworn in as the new president.

The promise is that within a year's time, parliamentary and presidential elections will be completed, with the army only protecting the process.

But the central question is how this predicament of division can be solved. I propose a plan of three specific points:

First, and immediately, a council should be formed to consider the latest constitution and the articles of disagreement. Within three months, the constitution should be amended and approved by the people in a referendum vote so the country can be united on future binding principles of society and the election processes that follow.

Second, the parliamentary election should precede the presidential election. Again, within another three months, Egyptians will know the political identity and spectrum of their parliament, and from this new election will emerge the majority and minority parties, inclusive of civil and Islamic parties.

Third, and most important to end the current polarization and violence, is the formation of a supreme presidential council, a board of trustees, made up of three groups. This council should have one-third representation by the civil parties, one-third representation by the Islamist parties, and one-third representation by wise men and women who are independent and not politically associated with any party.

Perhaps five to a maximum of 10 people from each group would be sufficient. This body would have the authority to discuss in depth the upcoming proposed changes to the Constitution and the proposed election process, and to vote on them. This way they are part of shaping the future; from the beginning they are included in the political process at the highest level. Disagreement may occur, but in the end, a vote among the members will be binding for all.

It is critical that during this transitional period the leadership does not repeat the past, and must deal with the Islamists fairly and inclusively. Also, Morsi must be treated justly in accordance with the law. Finally, the media must live up to the occasion. It must stop "reverse polarization" and work toward convergence, not divergence.

The Muslim Brotherhood and the Salifist parties are a real force in the Egyptian society. No civil, liberal government can succeed, even after new elections, if the Islamists are forced to work underground as a foe and the country remains divided. In fact, this is evident even after the June 30 event, when the Salifist al Nour party had a final say on who became the prime minister for the translational period.

Reducing Egypt's predicament only to the issue of a coup without realizing the central issues of division and violence in today's Egypt is too simplistic and dangerous. Every effort should be made to help build the new democratic nation with reconciliation and forgiveness, for the sake of Egypt and not for the benefit of one party or one group.

Ahmed Zewail is the 1999 Nobel Prize winner in chemistry and is a former US envoy to the Middle East.

© 2013 Global Viewpoint Network/Tribune Media Services

The New York Times

Published: 3 January 2013

Egypt's New Year Resolution

Op-Ed by Ahmed H. Zewail

CAIRO — Egypt is in turmoil, and many so-called experts have concluded that religion is the cause. It is not.

The source of turmoil is the gap between expectations of speedy change by those who made the people's revolution two years ago and the slow process of building an entirely new society.

Throughout my life I have never seen Egyptians expressing such an intense

Tyler Hicks/The New York Times

feeling of national ownership. This is one of the most important rewards of the revolution. The people are thirsty for real democracy after the revolution empowered them to seek their rights.

They may have been patient for 30 years of Hosni Mubarak's reign, yet are impatient with the progress made so far precisely because it is their own expectations they must live up to.

The liberals and protesters are fearful of a return to dictatorship. The Islamists who have suffered for decades from jailings, torture and underground struggle now have the above-ground legitimacy to govern and do not wish to go back to their unfavorable status.

Besides this political divide, there exists the silent majority, the so-called "*Kanaba* Party," or the couch party, which is sitting and waiting — and will vote when the opportunity arises — for the return of normal life.

One middle-class Egyptian Muslim, Ahmed Mostafa, told me, "A president from the Muslim Brotherhood is fine with me, but he should rule guided by our Islamic values and not in line with his *Gamaa* ideology," referring to the views of the Brotherhood.

"Only when he changes his mind from being a *Gamaa* Brother to a national leader can we have real progress in Egypt," Mostafa said, adding: "For now, to achieve the desired change we must shape the iron while it is hot. We are no longer fearful of our government."

After the presidential election in June of last year, Mohamed Morsi received the support of many Egyptians, including many who did not vote for him, in the hope of putting the country on the right track to development.

Egyptians wanted their elected president to succeed in addressing the real, numerous issues facing the nation, including the stagnant economy and the reforms needed in education and health care. If religion had been the dividing force, this support for the president would not have materialized.

However, after his decrees on a speedy ratification of the constitution and on limits to judicial rulings, opposition escalated. Islamist supporters of the president responded by organizing rallies of hundreds of thousands.

The organization of society along the lines of Islamists, liberals and a silent majority is not much different from what exists in established democracies. What is new and different for Egyptians is that the fear has disappeared and has been replaced with a sense of the power to shape their collective destiny.

What might be the path forward in the new year?

First, and most important, we need dialogue among the different parties. This will come about when the population feels that it is protected by the constitution and an independent judiciary.

Second, the government must select some major projects that would fulfill Morsi's campaign promises to revitalize the economy. I believe it is crucial to increase productivity, and this can only be achieved by improving education and a knowledge-based economy.

Third, everyone must accept that Egyptians are a religious people. Secularism will not work in Egypt any more than theocracy. What will work is governance that is guided by the Islamic values of the majority with protection of the minority rights.

In a different form, this structure, with a well-accepted constitution based on the principles of human rights and religious freedom, would not be too different from the situation in the United States, whose values are guided by the Christian faith.

Egypt has great potential because of the latent power of its human capital. We need to grab hold of the future now. This should be Egypt's New Year resolution.

Ahmed Zewail, a professor of chemistry and physics at the California Institute of Technology, was awarded the 1999 Nobel Prize in Chemistry.

This article has been revised to reflect the following correction:

Correction: January 7, 2013

Because of an editing error, an earlier version of this article incorrectly associated President Mohamed Morsi's ''Gamaa ideology'' with Al Gamaa al-Islamiyya, an extremist Islamic organization. In this instance, ''Gamaa'' refers to the Muslim Brotherhood, the president's former political organization, not Al Gamaa al-Islamiyya.

THE HUFFINGTON POST

TOP NEWS AND OPINION

Published: 28 November 2012

Syria: Is the World Waiting for Genocide?

Op-Ed by Ahmed H. Zewail

Cairo — From Nazi Germany to Rwanda, some of the most inhumane atrocities and genocides were committed while the rest of the world was watching. Today we are all witnessing atrocities and mass destruction in Syria. Again we are observing it all unfold before our eyes, with heavy hearts perhaps, but no effective intervention to stop it.

The regime of Bashar al-Assad is brutal. It is fighting a war with its own people, shelling them from the sea, bombing them from the air and murdering them in their own homes. According to the Syrian Observatory for Human Rights, more than 40,000 people have so far been killed. More than a million have fled the country, with hundreds of thousands becoming refugees in the bordering countries Turkey, Jordan, Iraq and Lebanon.

The regime is also extending the destruction to cities of immense historic value. All major cities have been bombed, from Aleppo in the north to Daraa in the south, Homs in the center, and Deir Ezzor in the east.

The situation in Syria today is no longer a political uprising bringing up the tail end of the Arab Spring. Rather, it has become a humanitarian calamity growing bloodier by the day.

We can't afford to stand numb, paralyzed under the cloak of excuses such as "The situation is very complex" or "The regime will ultimately fall" or the one promoted by those whose only interest is keeping the regime in power: "We must allow for diplomatic solutions."

The mission of UN envoys Kofi Annan and Akhdar al-Ibrahimi failed. The Arab League resolutions are ineffective. The UN resolutions are fragmented by the split among the veto-empowered members of the Security Council.

Some leaders think time will solve the problem. Their hope is that Assad's regime will ultimately fall from the heavy toll of the horrors it has spawned. From past experience with such regimes, this scenario is unlikely to happen. Assad, who belongs to the Alawite sect, a minority population in Syria, is still seeking to rule the whole population with an iron fist, including the majority Sunni Muslims. Everything Assad has done so far suggests he will do what it takes to survive, even using chemical weapons in the end. His regime is no different from that of his father, Hafez al-Assad, who leveled and poisoned a whole city, Hama, killing tens of thousands.

Syria is the proud heir of an ancient civilization that has a unique spectrum of minorities that encompasses Muslims and Christians of various denominations. There are at least ten such ethnic and religious groups. For centuries they all lived together peacefully. Now, with the internal war intensifying, this unity is dissolving into a civil and tribal war that not only will end Syria's nationhood, but will also spill over to the rest of the Middle East.

There is little doubt that an unstable Syria will destabilize the whole Middle East. The sectarian turmoil resulting from this tragic conflict will become contagious in neighboring Iraq and Lebanon. The latest event of a car bombing that killed innocent civilians in Lebanon is a warning sign for the looming spillover.

With Assad using the Golan Heights as a "card of fear," Israel may step in. This, in turn, will incite Iran, Hezbollah and possibly Turkey and Egypt to become involved. Already reports indicate that armed groups, including those that are linked to al Qaeda, are finding Syria a fertile ground.

Even China is coming to recognize that chaos in a country that possesses thousands of missiles and huge stockpiles of chemical weapons threatens not only the stability of the Middle East but also the flow of oil to the West and East alike.

In this case, the humanitarian cause and the global interests are indeed aligned.

Syria may appear to be a small country, but it is just the type of entagled conflict that can lead to a world catastrophe. It does not take much imagination to see Syria as the Sarajevo of the 21st century, leading to world war.

The world must act now and collectively. If we have the will, we will find a way. The least we can do immediately is to provide shelter and tangible aid to those fleeing across the borders. In parallel, the superpowers must support on the ground the Syrian free army and impose a no-fly zone, which will cripple the capability of this heartless regime to bomb innocent men and women.

With the Christian holidays weeks away, it would be shameful to celebrate the birth of a man of peace and humanity — whose native language, Aramaic, is still spoken in Syria's Ma'loula — when genocide instead of brotherly love stalks the very lands he and his disciples roamed millennia ago.

Ahmed Zewail, Nobel laureate for Chemistry; Special Presidential Envoy for Science to the Middle East.

The New York Times

Published: 20 May 2012 (Sunday Edition)

Egypt's March Toward Democracy

Op-Ed by Ahmed H. Zewail

CAIRO — A few days ago, I watched a debate between Amr Moussa and Abdel Moneim Aboul Fotouh, two of the leading candidates among the 13 running for president of Egypt. This stunning debate went on for more than four hours and was watched by millions of Egyptians and other Arabs. Contrary to the perception around the world that Egypt is inexorably sinking into chaos and intolerance, this debate in many ways reflects the hope for a new Egypt following last year's revolution.

Amr Nabil, Associated Press

From the time of Ramses II, the strong pharaoh who ruled Egypt thousands of years ago, until last year when Hosni Mubarak's reign ended, Egyptians were never able to witness a debate over who should take over the democratic reins in the highest office of the land. Our new culture of debate, together with the election of the Parliament last December, are milestones in the history of the nation, paving a new, but rocky, path toward democracy.

Unlike in nearby Syria or earlier in Libya, the Egyptian Army has taken the high road and protected the revolution in its infancy. And it has been the guardian of these unprecedented transparent elections.

The challenges facing the country, of course, are still monumental.

Among the most serious problems are economic hardship, the uncertainty of the political climate and the deterioration of security — a feature that Egyptian society faces anew. These problems have been compounded over the past 15 months as each of the three main constituencies involved in the revolution — the Supreme Council of the Armed Forces

(SCAF), which is in charge of the transition period; the politically liberal as well as Islamic-oriented parties; and the youth who triggered the uprising — have stumbled in one way or another.

Even some of the people most thirsty for transforming Mubarak's stagnant nation from a democracy-deficient to a democracy-rich society have, in despair, been yearning for the old stability.

True, there are chaotic symptoms — such as the conflicts among the different political parties and the occasional clashes between the SCAF, the Parliament and the government — but this is a form of the "creative chaos," in the words of Condoleezza Rice, that is a consequence of revolutionary changes that ultimately will lead to a stable democracy.

The recent French election is a lesson for us in the peaceful, civilized transfer of power. Looking back, we know the French Revolution some two centuries ago, through a liberation movement not unlike ours, was accompanied by widespread bloodshed and ugly political conflicts for many years.

It is a hopeful sign indeed that we Egyptians are still marching forward toward democracy with relatively little bloodshed. All signs indicate that a counterrevolution is not in store for Egypt. We will not turn back to a totalitarian governing system.

Perhaps the most encouraging of all is the confidence of Egyptians in their future.

The governor of the central bank of Egypt, Farouk Al-Okda, recently informed me that the hard-currency revenues coming from expatriates are the highest ever. Even the many strikes for betterment of education, improvement of health care and increase in salaries can be read as acts of high expectations for the future.

A rise in violence between some Muslims and Christians is cause for concern. But its origin and intensity are exaggerated in the media.

Egypt's Christian history is part of the fabric of the society. Egypt does not have a ghetto for its minority population nor segregation of students at schools, but indeed has some solvable problems to address, including those of civic society issues and representation in governance.

Growing up in Egypt, I witnessed the harmony between the peoples of the two of faiths. Together we celebrated Eid al-Fitr, Easter and Christmas, and together we lived in the same buildings and went to the same schools. The late Pope Shenouda III used to say: "We do not live in Egypt, but Egypt lives within us."

The current grand imam and sheikh of Al-Azhar, Ahmed el-Tayeb, has signed on a new constitutional paper demanding unity and human rights for all Egyptians.

In the post-revolution period, some bad actors, including those from the previous regime, seek to fan inter-religious violence in order to destabilize our infant democracy. The fact that it hasn't taken a deep hold is yet another sign of hope.

Naturally, the role of religion in politics is now being debated, and in fact the recent debate is telling of this change.

Dr. Aboul Fotouh was a member of the Muslim Brotherhood group that was established in 1928, and suffered from political persecution. He calls himself a liberal Islamist.

Mr. Moussa, on the other hand, who was a foreign minister and secretary general of the Arab League during the Mubarak era, stresses his experience and paints his opponents as religious extremists.

The open debate between the secular and religious orientations of politics was unthinkable over the past 60 years. This new openness means the Egyptian body politic is maturing.

Citizens are taking responsibility for their own fate by insisting that diverse visions and ideologies compete. In the end, Egyptians know that, for the first time, they can choose their future. It won't be dictated or imposed by anyone.

From my involvement in Egypt, I am confident that the SCAF will hand over the power to the elected president. I, however, believe that the SCAF wishes to have a "respectable exit" and some guarantees regarding the status of the army in the constitution of the new Egypt.

My message to the Egyptian people, and especially to the politicians, is simple: For the sake of Egypt, unite together to complete the passage from fallen dictatorship to emergent democracy by focusing on charting the new constitution.

No matter who comes to power, the constitution will protect citizens against abuse of authority either by the legislative or the executive branches. Luckily, Egypt still has a respected and robust judiciary system to complete the triad of democracy.

My concern is that the practice in Mubarak's era of "conflicts of trivialities" can cause the nation to drift away from the central issues of the constitution and economic productivity. The more effective this unproductive course, the longer the transition time to democracy.

It is imperative that we do not give up hope. The world must support a democracy that has passed its pregnancy stage and is now in the gestation period, ready for a new birth.

Ahmed Zewail was awarded the 1999 Nobel Prize in Chemistry. He is a professor of chemistry and physics at the California Institute of Technology and is playing an active role in Egypt's transformation to democracy.

© 2012 Global Viewpoint

Los Angeles Times

Published: 5 December 2011

Pillars of Change in Egypt

Egypt's election, like the revolution itself, is defying expectations.

Op-Ed by Ahmed H. Zewail

Egyptians are going to the polls to elect a democratic parliament, an experience they have not had for over half a century. This is an extraordinary and exhilarating event.

What's remarkable about it, among other things, is that only a week before the plebiscite began, an on-schedule election was thought to be impossible. The media were predicting that a fair election could not be pulled off and that, if voting did occur, it would be bloody and

Egypt's election, like the revolution itself, is defying expectations.

violent. But Egyptians weren't daunted. They remembered how, 10 months ago when the uprising began, many pundits predicted that Hosni Mubarak's regime was too strong for the revolution to succeed. Egyptians were not the kind of people to rise against their leader, they said, and if they did, they would have no idea how to create a democracy.

The election, like the revolution itself, is defying expectations. So far, more than 70% of eligible voters have participated, turning out despite long lines. Peace has prevailed.

So why have Egyptians gone against expectations, and what does the future hold in the postelection era?

When I was growing up in Egypt, we used to ponder the influence of the Nile's eternal nature on the character of Egyptians, on their peaceful, jovial and easygoing ways, and in some sense on their passive attitude toward change. Renowned Egyptian geographer Gamal Hemdan wrote of Egypt's character and its "genius of locus," the country's unique interplay between geography and culture. Egyptians, in general, have been tolerant and content, perhaps because time in their civilization has extended over millenniums, and changes in mere tens of years would seem abrupt. But as we are seeing once again, Egyptians are capable of sweeping and rapid change once it has been collectively deemed appropriate.

Since the 1880s, Egyptians have revolted four times, about once every 40 years. For the last half a century, though, change did not result in democratic gains, which has caused some to question whether that outcome would be possible this time. I believe democracy will prevail.

Five years ago, at the zenith of Mubarak's rule, I wrote about four pillars of change that could transform Egypt into a knowledge-based society and a democratic state: reform of the constitution, respect for the rule of law (including a strong and independent judiciary), a renaissance in education and reform of the media.

Since January, significant changes to the constitution have been made, the most important of which was limiting the presidential term to four years, with a two-term maximum. Although the constitution will not be finalized until early next year, it is clear that the power of the presidency will be limited and that it will be checked by an elected parliament and an independent judiciary.

There is also cause for optimism about the second pillar of change. During the Mubarak era, Egypt's respected and professional judiciary system was weakened, and in some cases manipulated, but the totality of it remained robust and capable of enforcing justice in the country, and it has been reinvigorated by the year's events. This was evident in the first round of elections last week, which was conducted completely under the supervision of Egyptian judges, with impressive results. Ultimately, with the establishment of democratic institutions, I believe the new system of the rule of law in Egypt will ensure freedom of speech and the equal rights of all Egyptians.

The third pillar of change — education — has already been inaugurated. During the Mubarak regime, a renaissance in education, which I have long advocated, was impossible to implement. But since the January revolution, the government has made great strides, among them passing a decree to establish the Zewail City of Science and Technology for the purpose of modernizing science and technology education and advancing the level of productivity. We have already raised nearly $200 million as the seed for an endowment to establish a campus in Zewail City that will be governed by a board of trustees made up of distinguished scholars, including the president of Caltech and six Nobel Prize winners. These steps are the beginning of a nationwide transformation.

The one pillar that has not yet been established is reform of Egyptian media. The media, both public and private, still lack depth and breadth in coverage. Media outlets tend to be partisan rather than objective, and some promote disunity. Because the media — especially television — have great influence on a population still plagued by illiteracy, it is important that journalism improve. It is inevitable that the influence of the first three pillars of change will ultimately lead to a new and healthier media for Egypt.

I believe Egypt's transformation is irreversible, that no force can ultimately deter the movement toward democracy. The "barrier of fear" has dissolved, and Egyptians know they can demand and institute change, and that the mind-set of Tahrir Square is lasting. Moreover, based on my personal meetings with the Supreme Council of the Armed Forces, I am convinced that the army will transfer power to a national civilian government.

In the past 10 months, there have been mistakes made, but the ongoing, judicially supervised elections demonstrate not only good intentions but also the desire of a broad swath of Egyptian society to have a free and democratic state. The Western fear that Islamists will take over and hijack democracy is exaggerated. Egyptians since the time of Akhenaton have been a people of faith, which has been a force for their unity in society. Islamists may well dominate the new parliament, but in a true democracy their performance on behalf of the people will be the key to their staying power.

Egypt's future challenges go beyond the transition to democracy. The country must contend also with its poor systems of education and healthcare, and perhaps most critical, it must address the vast economic gap between rich and poor. Domestic security must be strengthened, so that investment and tourism can reach their potential, and other economic reforms must be instituted to bring the kind of prosperity that Egyptians expect in the post-revolution era. And Egyptians must also find a way to forgive, even if they can't forget, those who hobbled the country's past aspirations.

In the coming years, Egyptians must capitalize on their "genius of locus" in order to move forward, unfettered by the past.

Ahmed Zewail is the 1999 Nobel laureate in chemistry. He is a professor at Caltech and serves on President Obama's Council of Advisors on Science and Technology, and is chairman of the board of the Zewail City of Science and Technology.

The New York Times

Published: 5 October 2011

As Elections Loom, Egyptians Must Unify

Op-Ed by Ahmed H. Zewail

CAIRO — "Where is Egypt going?" a driver named Mohamed asked me recently. It is the question on everyone's mind as the Arab Spring of popular revolution gives way to the new season of free elections this autumn.

At this unique moment in history, there are two critical challenges that face this nation at the heart of the Arab world. The first is how to further catalyze and consolidate the democratic transition through re-establishing unity among all Egyptians. The second is the related issue of achieving a commitment to peace in the Middle East that is genuinely supported by the Egyptian public.

In the months since Hosni Mubarak's ouster the road to democracy has been rocky, but the spirit of optimism is still high. I remember the thousands of people who lined up across Tahrir Square at the entrance to the American University in Cairo to witness the announcement of the National Project for Scientific Renaissance and the building of the new City of Science and Technology. At last, it was felt, Egypt would rejoin the future that had been blocked by dictatorship. My nationally televised speech on that occasion was entitled "*Musr al-Amal*" ("Egypt: The Hope").

That optimism was shaken by an event that took place on the last Friday of July, when the world witnessed the largest demonstration since the "de-throning" of Mubarak. The Islamists who had gathered called for the new Egypt to be governed according to strict Shariah law. The display of Saudi-like flags among the crowd prompted headlines such as "Bin Laden is in Tahrir." Then, in September, came the attack on the Israeli Embassy in Cairo in response to the killing of Egyptian servicemen by Israeli forces in Sinai. This caused great alarm, especially in the United States, about the security of Israel and the future of peace in the Middle East.

Despite these troubling events, I remain optimistic about Egypt's future. The Egyptians are no longer fearful of their rulers, they know how to demonstrate, and they are determined to change governance. But certain guiding tenets must now be followed.

First, Egyptian society must focus on its long-term goals. It is unconscionable for the media to continue with its shallow methods of the past. The country needs to have a constructive discourse on fundamental issues, such as establishing constitutional principles on religion and governance, revamping the education system and boosting the stalled economy.

Second, it would be a mistake to unjustly alienate all people who were associated with the fallen regime. They should be regarded as fellow citizens whose resources and energy should be redirected toward building the future. Egypt cannot afford to have the vital energy of its active intellectuals consumed by the past or its political leaders absorbed by concerns over what slice they will get from the cake of revolution.

Securing democracy will require that the significant divergence of ideologies and political alignments that have emerged in recent months be once again put aside to fulfill the common aim that first united the people: the fall of the system, or "*iskat al-nizam.*"

Lastly, but most importantly, the army and the government must chart a clear road map for the weeks and months ahead.

The Egyptian public still highly respects the military, but people are wary of the fact that the Supreme Council of the Armed Forces remains the main political entity governing the transition to democracy. The fact that the Supreme Council is dragging its feet unsettles everyone who recalls how the machinations of the previous regime frustrated democratic aspirations.

The best cure for this suspicion is clarity and communication from the Supreme Council on a time table for the coming elections and its stance on a number of issues: the emergency law, the military courts and the voting of Egyptians abroad. The council must also specify how and when the new government will be put in place. The massive demonstrations in Tahrir Square are continuing. In order to push forward, the country needs stability and security.

One way to bridge the trust gap with the public might be through the establishment of a council of eminent citizens who can mediate between civil society and the governing regime as the transition takes place. Such a "roundtable" between Solidarity and the military rulers of Poland guided the transition to democracy there toward the end of the Cold War.

The present confusion and distrust threatens the advance of the economic revolution upon which success or failure of the democratic revolution ultimately rests. For this reason alone we must make the political transition properly and as soon as possible.

It would be a mistake to conclude that the current strain between Egypt and Israel will mean the end of the peace treaty between the two countries. During the height of the revolution there were no banners against Israel in Tahrir Square. The current Israeli ambassador has repeatedly said that before the recent riots at the embassy he was never mistreated by the people. What ignited the anger of the young people who attacked the embassy was the killing of the servicemen on Egyptian territory in Sinai and statements of support for Mubarak as "the best friend of Israel" by officials in the Netanyahu government.

Since the revolution, Egyptians feel that "we the people," not the rulers, must decide what kind of relationship there will be with Israel.

Following Tunisia, Egypt and Libya, the whole Middle East is awakening. The spirit of the Arab Spring will extend to the Palestinians, who will demand the end of occupation just as Egyptians demanded the end of dictatorship.

The major powers should not position themselves against the current of history. Instead, they should commit to ending the Palestinian-Israeli conflict through the creation of a new state based on the pre-1967 borders. Such a course is in the best interests of the world community because it will shift the energies of the Arab people toward building their own states instead of perennially focusing on the conflict with Israel. At this decisive moment, whoever uses the Palestinian-Israeli conflict for political gain is guilty of a crime against hope.

The highest priority is to prevent the hopes and aspirations of the Arab Spring from being derailed by disunity or the manipulation of the heated emotions that have long swirled around the central issue of peace in the Middle East.

Ahmed Zewail is the 1999 Nobel Prize laureate in chemistry, and the first U.S. Envoy to the Middle East.

HUFFPOST WORLD

Published: 22 June 2011

A Compass of Hope for Egypt: The New "City for Science & Technology" Is the Aswan Dam for the 21st Century

Op-Ed by Ahmed H. Zewail

Cairo — Nearly 100 days after the revolution, Egypt is very different from the country I experienced when millions were on the streets calling for the fall of Mr. Mubarak's regime. Despite a myriad of problems, now there is a new energy, or, as the Egyptians say "*hawa gadid*" — a new air. The big question is how to channel this energy to forge a new Egypt that is democratic and sustainable both politically and economically.

The key to moving forward is building confidence among the people with an immediate high-profile project that captures their imagination and symbolizes what the future can bring. Just as the Aswan Dam did that for an earlier generation, the new "City for Science and Technology" now underway can do for today's hopeful youth.

In the 1960s, I personally lived the resounding impact of President Nasser's vision of constructing Aswan's High Dam as a "national project" for controlling the Nile irrigation and the production of electricity. As the young journalist Emad Ahmed wrote in a recent essay on "Egypt's Bridges" to the future, the post-revolution national project for Egypt comparable to the Aswan Dam must be education.

Every family in Egypt understands this. They have personally experienced the deteriorating education system over the past 30 years of Mr. Mubarak's reign.

Especially for the "Youth of Facebook" who ignited the revolution, the focus on a breakthrough in education that can bring Egypt back to world-class status is in accord with the principles and spirit of their movement — which they fear could be overtaken by "politics as usual" rooted in the past.

As Emad has written, two dominant visions have shaped the Egyptian political imagination over the past 60 years. The first has been the socialist party or "*al-Hisb al-Ishtraki*" which came with Nasser's 1952 "revolution". To today's youth that vision represents the past.

At the moment, the most organized force is the Muslim Brotherhood, or *"Akhwan al-Muslimin"*. For Emad, they represent the transitional present. From a historical perspective, the *Akhwan* also are part of the past as they were founded in 1928, even before Nasser's time. Their appeal comes mainly as a result of their effective religious message and organized charity work, and because they resisted the regime for so long.

The youth movement is aware that old visions can not take Egypt into the future. So, in the months since Mr. Mubarak was overthrown with the Army's admirable support, the youth, along with a broad spectrum of ordinary Egyptians, have kept that spirit alive by continuing to go to Tahrir Square on Fridays in what is called *"millioniah"* or gathering of a million people. They coined a name for each gathering — a Friday of change (*takhier*), of anger (*khadab*), of correction (*tas'hih*). Their demand is that the road to democracy be paved through the establishment of proper constitution, elimination of old-regime influence, and achievement of justice and equality. Their expectation is a quick remedy to a better economic status.

After so many years of inertia and dictatorship, however, the reality is that these changes will take years. In the meantime, the people need a compass of hope that unites the country and instills confidence and pride.

On June 3, a totally different Friday dawned on the country. It was a "Friday of hope" for Egyptians. The day before, a national campaign was launched to build the new City of Science and Technology, following the unanimously-approved legislation by the Cabinet of Ministers, and a decree of support from the Supreme Council of the Armed Forces.

This "city of the future," as it is already being called, which is being built on 300 acres on the outskirts of Cairo, has a transparent governance structure and is completely independent from government regulation. The board of trustees that has already been formed includes six Nobel Laureates, the current president of Caltech and former president of MIT, and a number of influential Egyptians such as Mohamed El-Erian, CEO of PIMCO, who has already made a large personal donation to the city. Sir Magdi Yacoub, the renowned heart surgeon based at Imperial College in London is also a member.

It is not surprising that the project has been enthusiastically embraced by public opinion. Ibrahim Issa, a prominent journalist and one of the leaders of the Tahrir protests, has said, "It is the only important thing proclaimed since the revolution." Ahmed Moslemany, a popular TV commentator, has announced to millions of his viewers "it is the only way to the modern world".

For twelve years, since I was awarded the Nobel Prize, I have been laboring to get this project off the ground, only to be frustrated by bureaucrats lacking vision and Mr. Mubarak's indifference. The "new air" of the revolution has breathed new life into the project.

Our goal is to develop a non-profit institution of higher learning that is merit-based (no "wasta" or connection), and our model is a hybrid of Caltech, an institution I am familiar with for more than 30 years, the Max Planck Institutes, and Turkey's Tech. Park. The objective is to revive the production of new knowledge by Arabs and to bring the advances of science and technology to the market and society in this Arab awakening epoch.

Our aim is to demonstrate that "Egypt can." This, by itself, will have a huge impact on regaining national pride.

Even with the present economic hardship, Egyptians have decided to invest in the future, with billions of dollars in land and buildings for the project. In weeks we have already nearly collected the first $100 million in our campaign for a $2 billion endowment that will ensure the long-term success and independence of the project.

Our hope is that the international community — the Gulf States, the G-20 and the G-8 (which pledged $20 billion at the Deauville Summit) — will create a genuine partnership with Egypt to invest in the education of our youth so that the gains of the revolution can be consolidated with benefits to the region and the world.

The benefits for all are clear if this region that is so important to the world can at last make progress and develop.

When the people of Egypt fulfill their dream of democracy and sustained growth, it will go a long way toward opening the broader MENA market of close to 400M people for business.

Investment in education and economic prosperity is the best way to cure fanaticism and for establishing a just peace in the Middle East. An institution such as the City of Science and Technology will surely be a center of enlightenment and global cooperation.

The Egyptian revolution, which had no ideology but peaceful change, demonstrated clearly that the assertions that Muslims and Arabs are incapable of participating in the modern world or that they are in violent conflict with Western civilization were unfounded. Like everywhere in the world, people of the Middle East aspire to liberty and justice. They wish to have a better life and a decent education for their children.

After more than 50 years of supporting undemocratic autocracy in the region, nothing would more successfully win the hearts and minds of Egyptians than real support for this tangible bridge to the future for a people who have liberated themselves with dignity and civility.

Ahmed Zewail, Nobel laureate for Chemistry; Special Presidential Envoy for Science to the Middle East.

FINANCIAL TIMES

Published: 25 April 2011

Fund Egypt's Future to Save the Arab Uprising

Op-Ed by Ahmed H. Zewail

As I was leaving Cairo after Hosni Mubarak stepped down, I asked Esraa, a young woman who was one of the leaders of the revolution: "What was your objective?" She said, "*taghier al nezam*", a change of the system. The Egyptians brought down the head of the system, but not the system itself. That is the challenge now.

Egypt's revolution, like Tunisia's, represents a model for change in the Middle East. These societies are not fragmented by tribal or sectarian conflicts. Despite differences of faith or even the occasional collisions between them, Egypt is united. In contrast, the second model for revolutions is that of Yemen, Libya and others in the making. In these cases, unfortunately, tribal and sectarian conflicts may lead to chaos and civil war, ultimately dragging the Middle East backward into conflict and fanaticism, not forward.

To avert this, the revolutions in Egypt and Tunisia cannot be allowed to fail. Egypt is the key. With 85m people, it is the largest country in the region and the heart of the Arab world. Making sure it succeeds is essential for the spread of democracy and stability of the world's energy supply as well as for peace in the region. What can be done?

From my experience pushing for reform over two decades and as a negotiator with the youth and government leaders during the revolution, I know what in the long term is needed most — a decent education system. The so-called "children of Facebook" who fomented the revolution know Egypt was once ahead of South Korea in scientific research and development. They know that in the 30 years Mr Mubarak sat in his palace and Egypt deteriorated, China has lifted hundreds of millions out of poverty, sent astronauts into space, built megacities and high-speed trains, and brought its students up to world standards. They ask why Egypt cannot do the same thing.

This will, of course, take time but it is imperative to begin now. Of Egypt's many problems the three most urgent to address are governance, economy and education. The army's Supreme Council, now the ruling political entity, has to ensure swift political changes. Egypt badly needs national unity and reconciliation. But to take the critical long-term steps to transform society it needs financial support. While Egyptians themselves must fashion the new nation, they need help in rebuilding sustainable institutions. The place to start is with the pivotal project, "renaissance in education and development", whose acronym is the first command of the Koran — read.

For years the west supported Mubarak and gave aid for what it hoped was stability — but was actually stagnation — in the Middle East. What Egypt needs now is a global partnership of private and government organisations to establish a fund to finance a revolution in education. This should be directed by a board of trustees from renowned Egyptians and world leaders in co-operation with the Egyptian government.

Such an effort would need an initial $1bn from private and government sources. Further funds then could come from other nations and be deployed by the World Bank, the Arab Bank and the Islamic Development Fund. Repudiation of debt will redirect national resources to this and other vital projects. Egypt does not possess rich natural resources. Its agricultural area is relatively small — less than 10 per cent of the total land. Its growth relies on tourism, Suez Canal tariffs and foreign investment. Yet Egypt is rich with human capital. According to the United Nations, Egypt's population will grow to 114m before it stabilises in the year 2065.

The psychological influence of launching such a fund at this moment cannot be underestimated. Like Tunisia, Egypt today is in flux, as a range of interest groups contend over the direction of the future. By showing that effective, and youth-based, institutions can be built, this project would point Egypt in the right direction.

In Egypt, every family is suffering from the deteriorated schooling and university system of the Mubarak regime. What families want most of all is to secure a good education for their children.

It is in the best interests of everyone — the Chinese, the Americans, the Europeans and the other Arab states — who wants long-term stability in the Middle East that the peaceful democratic revolutions in Egypt and elsewhere succeed. Time is of the essence!

The writer was awarded the Nobel Prize for chemistry in 1999. Currently he is a professor at the California Institute of Technology, serves on President Barack Obama's Council of Advisors on Science and Technology, and is involved in Egypt's transition to democracy.

THE ⟐ TIMES

Published: 21 February 2011

Education System Needs Its Own Revolution to Succeed

Op-Ed by Ahmed H. Zewail

The process of transformation begins with democracy, but it does not end there. The first uprising brings political change; a second is now needed to transform Arab learning. The failure of Arab education is one of the underlying causes of youth discontent in the region and has serious cultural, economic and political consequences.

Today, the Arab world's contribution to international scientific research is negligible. No Arab university regularly ranks among the world's 500 best institutions. A major project to eliminate illiteracy must be a top priority of Arab governments. I find it hard to believe that between 25 and 50 per cent of the Arab population of 350 million remains illiterate. Egypt's higher education system puts hundreds of thousands of students through a university education that is not sufficient for our modern world. A hierarchy of universities must be created. The best operate as non-profit organisations. Technical and vocational education must be upgraded to increase skill levels.

Finally, enhancing scientific research is essential for future development. Arab nations should allocate substantial resources to establish research and development centres of excellence.

The current education system has suffered, like every other area of Egyptian life, from corruption. It is time for meritocracy to replace nepotism and favouritism.

The choice is our eithers: either to become 350 million people of the cave, *ahl al-kahf*, or 350 million people of the cosmos, *ahl al-kawn*.

Ahmed Zewail won the Nobel Prize for Chemistry in 1999 and is the US science envoy to the Middle East. This article is abridged from Reflections on Arab Renaissance in the Cairo Review of Global Affairs.

THE ☙ TIMES

Published: 16 February 2011

We Must Unleash the Power of Egypt's Youth

This was an ideology-free revolution. Politics must not stand in the way of progress.

Op-Ed by Ahmed H. Zewail

The Egyptian people have overthrown the Mubarak regime in a peaceful revolution. Now that the tumult has subsided, the hard work of reconstruction must begin. There is a strange mix of excitement and trepidation in the air, but underlying it all is the prospect of real progress— not least in reintroducing Egypt's leadership of the Arab world.

For democracy actually to succeed, however, concrete steps must now be taken. First of all, the army, as it has pledged, must swiftly transfer authority to a civil government and step back into its role as guardian of the Constitution. Using Egypt's respected judicial system, a new constitution will be written to incorporate the ideals of liberty, justice and national unity.

A brutal regime was brought down by the peaceful power of technology.

Equally important, Egypt's new leaders must develop a national vision for economic and social development. The birth of democracy in Egypt requires that we sacrifice ideological battles and political games to focus on genuine reforms. Our young people cannot afford to be lured into traditional politics as usual.

The country faces big problems. At a national level, Egypt has to cleanse the deep-rooted corruption of the old regime and begin immediately to reform the bureaucracy of government.

Under the Mubarak regime, tourism, the Suez Canal and natural resources made up the bulk of the country's wealth. That is too narrow an economic base for an ambitious 21st century country. Egypt is rich in human capital but it suffers from appallingly low levels of literacy. With 85 million people, nearly half of them on the poverty line, the number 1 issue for reform is education. The country will never compete in global markets while 30 per cent of its population are unable to read and write.

The ideals and aspirations of Egypt's young people are real. What they are seeking is a new Egypt. They know their glorious past and wish to forge a new future. Their revolution was peaceful, civil and technologically advanced. They used the internet, texting, Facebook and other systems to co-ordinate their activities and ensure peaceful action against the brutal force of the regime.

The revolution did not have a hero of the Left or the Right. Nor did it have slogans such as "Death to America", "War with Israel" or "Islam is the solution". Young people cleaned up the streets, organised the traffic and even made human shields to protect national treasures such as the Egyptian Museum.

President Obama acknowledged the nature of the revolution when he declared from the White House: "Egyptians have inspired us, and they've done so by putting the lie to the idea that justice is best gained through violence." Egypt's youth — in contrast to the stereotype of young Muslims as the source of terrorism — deserve to be honoured.

Perhaps the most fitting way would be for the United States, Europe, Japan and China to create a significant fund to invest in their education. What we need most in this part of the world are science and technology centres of excellence. Such institutions would become beacons of knowledge, boost national pride and enable our youth to participate in the global economy.

One of the forces that will drive progress in Egypt is an appreciation of the value of scientific thinking. Some 35 per cent of the Egyptian population are 14 years old or younger. *It had no hero of Left or Right, no* We have seen how India's investment in science *slogans saying 'War with Israel'.* has paid off, propelling the country up the global league of nations. If we instil this core value in our young people, we will create a powerful workforce that can truly change Egypt.

This opportunity must not be lost. In the 1960s, with the aid of the former Soviet Union, Egypt constructed the Aswan High Dam, a national project that generated huge power for Egypt's economic development. Through their peaceful revolution, young people have generated energy on a similar scale for Egypt's human development.

Egypt is the heart of the Arab world and the beating of that heart is being felt across it. It is more critical than ever that we support and sustain the spread of liberty and democracy in the Middle East. If they are to endure in this region, we need to train the minds of the generation who made this revolution happen. Joining hearts and minds together is what will ultimately make us part of the modern world.

Ahmed Zewail won the Nobel Prize for Chemistry in 1999 and is a professor of chemistry and physics at the California Institute of Technology. He has been in Egypt during the fall of President Mubarak's regime and is in Cairo to aid the transition to democracy.

The New York Times

Published: 2 February 2011

Egypt's Next Steps

Op-Ed by Ahmed H. Zewail

CAIRO — The revolt that has erupted across Egypt is in many ways historic and should take the nation into a hopeful future. What's unexpected, even by the Egyptians themselves, is that this intifada is led by youth, the so-called Facebook children, with no religious or ideological agenda other than a better future for Egypt and its people.

In this difficult time, the military has earned the expected respect of the masses by acting professionally to maintain safety and stability as the guardian of the Egyptian people. By reclaiming the future while maintaining stability, these two forces of the youth and the military offer great hope for an orderly transition to a new Egypt.

Clearly, it is time for fundamental change in Egypt, not just cosmetic alterations. There are several reasons for the current uprising that must be borne in mind in order to figure out where to go from here. The people of Egypt have finally lost patience with power games among those surrounding the president over succession to his son, Gamal Mubarak; the lack of transparency among those who held power; and the phony elections that led in the last Parliament to a majority by Mubarak's party, effectively with no opposition.

Though Egypt has seen some economic progress in recent years, the masses of the poor have been left behind, and the middle class has actually gone backward. Only the small elite at the top has benefited lavishly by exploiting its influence with the government. The corruption resulting from this marriage and the constant demands for bribes by officials has further exhausted the tolerance of the people.

Finally, the education system, which is central to every Egyptian household's hopes of progress, has deteriorated into a sad state that is far below Egypt's standing in the world. The system failed in a big way, especially when I compare it with the one I personally experienced as a student in Alexandria in the 1960s. Moreover, scientific research in Egypt, which was ahead of South Korea, has now fallen to the tail of global rankings over the 30 years of the regime's governance.

Where do we go from here?

There are four important steps that must be taken to resolve the current crisis:

First, a council of wise men and women should be assembled to map out a new national vision and draft a new constitution based on liberty, human rights and the orderly transfer of power.

Second, the independence of the judiciary must be guaranteed.

Third, free and fair elections must be conducted for the upper and lower houses of Parliament and for the presidency, overseen by the independent judiciary.

Fourth, a new transitional government of national unity must be formed as soon as possible.

Egypt is in a transition, and it is important that the Egyptian people realize that in the coming days solidarity will be a key force for a successful outcome. The role of the military must be to maintain order and to protect the people from looting and crimes, and not to interfere in the formation of the unity government. Longstanding political parties and organizations should for now put aside their own agendas and place their priorities on building a stable bridge to Egypt's democratic future.

In order for this plan to succeed, President Hosni Mubarak must step down now. Mubarak came to power as a hero who fought bravely in Egypt's wars and headed the nation's air force. He can act heroically again if he leaves power immediately so that the transition to a new Egypt can take place in an orderly and peaceful manner.

Dr. Ahmed Zewail was awarded the Nobel Prize for chemistry in 1999. Currently he is a professor at the California Institute of Technology, serves on President Obama's Council of Advisors on Science and Technology, and is the president's special envoy for science to the Middle East.

© 2011 Global Viewpoint/Tribune Media Services

Herald INTERNATIONAL Tribune

Published: 30 September 2009

Obama's Sweet Egyptian Date

Op-Ed by Ahmed H. Zewail

In August, I returned to Egypt, the country of my birth, for the first time since President Obama spoke in June at the University of Cairo. I discussed the president's address with a veteran Egyptian diplomat, who described its impact as "historic." Obama's words were regarded as a momentous break from the past, spoken by an American president who respects Muslim faith and culture, and is optimistic about future relations with Muslim nations.

During that visit, I also got — literally — a taste of how the speech had gone down in the proverbial Arab Street, or at least in the streets of Cairo. The city's markets were piled high with dates in preparation for Ramadan, which observant Muslims mark with daily fasts, often broken with a date at sunset. Different varieties of the fruit are named after popular and unpopular personalities, depending on quality, and priced accordingly. This year, in the Cairo markets, the "Obama date" was of the best quality and commanded the highest price.

Here was a hopeful — and sweet — response to the president's overtures in Egypt. However, my friend spoke of a general sense that the next move too was up to Obama. I found this attitude profoundly unsatisfying. While Obama may be in a unique position to catalyze progress on the Arab-Israeli front, beyond that the impetus for change can only come from the Muslims themselves.

Egypt should take its cue from the fact that the president chose to speak not before government officials, but before an audience of educated young people. As a scientist educated in both Egypt and America, I appreciate the president's call for new education and science partnerships between Muslim nations and the West, for it is these areas that have the greatest potential to move a society forward.

This idea is by no means foreign to either Egypt or Islam. I came of age in Egypt after the Gamal Abdel Nasser revolution. Most Westerners today are unaware of the extent to which Nasser's regime promoted education as the vital engine of progress. I was one of many young Egyptians who reaped the benefits, receiving an excellent public-school education in a system that encouraged women to attend college with men and enabled many Coptic Christians to be prominent teachers and professors.

The religion of my youth did not advocate intolerance of other faiths, nor did it interfere with freedom of thought. I was raised in a devout family, but when my friends and I met up at the mosque, it was often to discuss academic subjects or to socialize. When we heard the word jihad, it meant to us ijthad, or "to strive," to excel personally and academically.

This was the environment that propelled me to enroll at the University of Alexandria, excel in my science studies with outstanding professors, win a scholarship for graduate study at the University of Pennsylvania, and become a professor at the California Institute of Technology.

My experience was hardly unique, and it offers an instructive example of what a commitment to education, unfettered by religious orthodoxies, can accomplish. When optimists speak of cultivating a spirit of progress among Muslim nations, we need look no farther back than the Egypt of my childhood, when the country had the best universities and richest cultural milieu in the region, and was a center of secular and religious learning.

Today we see a similar dynamic at work alongside internal liberalization in countries as ethnically and geographically diverse as Turkey, Malaysia, Indonesia, and some of the Gulf States, where substantial resources are poured into building up educational and civic infrastructures and working to overcome stereotypes that the Islamic world is hopelessly mired in oppression, religious extremism, and sectarian conflict.

Today in Egypt and throughout much of the Muslim world, more than a quarter of the population is under the age of 30. Neither Muslim governments nor the West can afford to overlook this immense reservoir of human talent and potential.

The most effective way to tap it is through a revitalized educational system, from the elementary grades through college, that is willing to capitalize on the best of both Muslim and Western traditions of learning, with a new emphasis on science and technology, and a recognition that assimilating new ideas represents not a departure from an authentic Muslim heritage but a return to it. A sustained investment in education is what will ultimately lead to greater economic prosperity, enhanced quality of life, and true democratic reform.

It is essential that Muslims embrace the American president's offer to collaborate in establishing regional "centers of excellence," among other initiatives in higher education. Egypt in particular should lead the way to advance educational and government reform. Otherwise, despite its auspicious start, Obama's Ramadan "date" will remain little more than a passing encounter, with limited prospects for a closer relationship.

Ahmed Zewail, a professor at Caltech, received the Nobel Prize in Chemistry in 1999. He was recently appointed to President Obama's Council of Advisors on Science and Technology.

THE INDEPENDENT

Published: 24 October 2006

The West and Islam Need Not Be in Conflict

We must not create barriers through concepts such as 'clash of civilisations'.

Op-Ed by Ahmed H. Zewail

Five years after September 11, we must ask, can western wars solve the so-called global conflict with the Islamic world? The answer, in my opinion, is no. A far better state of world peace would be achieved if the West would make a serious commitment to the just resolution of conflicts, and be genuinely involved, using a fraction of war costs, in building bridges to progress and peace with an understanding of the profound role of pride and faith in the lives of Muslims.

The vast majority of Muslims are moderates working for a better future and seeking a peaceful life. As evidenced by past achievements, Islam in its pristine state is not a source of backwardness and violence. As recently as the September 11 event, the majority of Muslims were, as the rest of the world was, against its violence. However, if despair and humiliation continue in the population of more than one billion Muslims, the world will face increasing risks of conflicts and wars.

As a cultural product of both "East" and "West", I do not believe there is a fundamental basis for a clash of civilisations, or that the West is the cause of all problems. Muslims are ultimately responsible for their plight. But the West has been more reactive than proactive toward the Muslim and Arab world, and has yet to implement a sustainable and equitable policy. For at least half a century Arabs have witnessed inconsistency in foreign policy, support of undemocratic regimes for the sake of securing resources and influence, and insensitivity to their culture and faith.

Here, I would identify four guiding principles for a new perspective. The first, and essential, point is political. The West in general and the US in particular should chart a vastly different foreign policy with the aim of gaining the confidence and cooperation of Muslims for solving complex conflicts.

In the Middle East, it is clear that peace will never be reached without solving the Israeli-Palestinian conflict. A two-state solution must be found and enforced. The unsettled conflicts in Lebanon and Iraq and with Syria and Iran call for solutions at the roots of the problems: occupation and borders; prisoners, refugees, and their right of return; and skewed international policy. Force and isolation will not solve these problems. Instead we need a comprehensive policy of fairness and firmness, perhaps established in an international conference and enforced by the United Nations.

Second, support for democracy in governance should be genuine. The West cannot and should not attempt to impose "Western democracy" and "Western values" by force on a culture proud of its heritage and faith. Many in the Muslim world admire the accomplishments and democratic values of the West, but people are mistrustful of "conditional democracy" and frightened of a culture now regrettably perceived to be of one of violence, sex, and other obscenities. Double standards and inconsistencies confuse people about Western intentions, and are used by totalitarian regimes to achieve their goals.

Third, foreign aid should be redirected toward economic development. Traditionally, an aid package is distributed to many projects, the major portion of which is for military support. The number of projects involved and the lack of an effective monitoring system, not to mention the influence of bureaucracy and corruption, results in few successes.

Directing aid toward the building of human capacity can be achieved through funding of innovative pilot programmes for enterprising individuals/groups in the free market, and invoking the expertise, and even the in-field labour, for the know-how. The use of aid programmes to support undemocratic regimes or groups is a grievous error.

Finally, education and research should be modernised through partnership. I see great opportunities for the people of the Muslim and Arab world, not less than those realised by China or South Korea. The West can help in the modernisation of education and research and development. I believe it is possible with the available talent and funding from rich Arab countries, and the know-how from the West and other world powers, to transform higher education.

Throughout history, people develop an interest in cultures and dialogues for the sake of mutual benefit. Even in one organ, the brain, 100 billion neurons work together to make a living human, and in our homes, cities, and countries we do the same. In an interdependent world, it is in the best interests of both the West and Muslim world to communicate through dialogues and to achieve global stability and mutual benefits from technology, commerce, energy, and cultures. We must not permit the creation of barriers through rhetorical concepts such as "clash of civilisations" or "conflict of religions", which are of no value to the future of our world.

The author is the only Arab Muslim to receive the Nobel Prize in science, 1999.

THE INDEPENDENT

Published: 24 August 2006

We Arabs Must Wage a New Form of Jihad

We must not be distracted by old ideologies and conspiracy theories.

Op-Ed by Ahmed H. Zewail

The cataclysmic wars in Lebanon, Palestine, and Iraq have uncovered the reality of Arab unity and plight, and the collective conscience of international society. It is abundantly clear that the Arab people must themselves build a new system for a new future. The current state, as judged by a low GDP, high level of illiteracy, and deteriorating performance in education and science, is neither in consonance with their hearts and minds nor does it provide for their political, economic, and educational aspirations.

Yet this is the same Arab world that produced leading civilisations, world-class universities, and renowned scholars and scientists. Clearly, something has gone seriously astray.

As someone from, and directly involved with, this part of the world, I am convinced Arabs are qualified to regain their glorious past. Arabs have two-thirds of "proved oil reserves", and copious sunlight for possible alternative energy. They have their own market, the potential for an Arab Union, and many Arab countries are strategically positioned, geographically and politically. The people have a unique culture of community and family values, and their faith is inclusive and pluralistic. Above all, the Arab world has people with talent and creativity, with nearly half of the population in its youth. These are forces for progress, but without nurturing intrinsic talent and establishing a cogent system of governance the status quo will prevail.

In my view, there are four "pillars of change" that would support an imperative historic renaissance for transforming the current state of affairs. First, a new political system must be established with, at its core, a constitution defining the democratic principles of human rights, freedom of speech, and governance through contested elections. A select delegation of honorable intellectuals, respected political personalities, and thoughtful religious scholars, perhaps under the patronage of supreme-court judges, should form a council to debate and chart a new constitution for a final referendum involving the people. The co-existence of religious values in the lives of individuals and secular rules in the governance of the state should be clearly defined. There is no need to fear conflict, as reason and faith are driving forces in western democratic societies and in some Muslim countries such as Turkey and Malaysia.

Second, the rule of law must in practice be applied to every individual, independent of caste, faith, or background. Currently, some rules of law are either unenforced or selectively enforced, resulting in demoralising practices. Besides being a prime cause of poor economic growth, poor governance breeds corruption which cripples investment, wastes resources, and diminishes confidence. If rules are applied fairly, people acquire security and faith in their system.

Third, the methods used in education, cultural practices, and scientific research must be revisited, reviewed, and revitalised. The goal should be to promote critical thinking and a value system of reasoning, discipline, and teamwork. The government should remain responsible for the primary education of all. Higher education should be based on quality not quantity, receive merit-based funding, and be free of unnecessary bureaucracy. Not the least of the benefits of educational reform is to foster the pride of achievement at national and international levels.

Fourth, an overhauling of the Arab media is necessary. Currently, there are numerous satellite TV channels and several so-called media cities generously financed, perhaps much more than research institutions. Yet people are inundated with mind-numbing and propaganda programmes. The conceptually new al-Jazeera has become a very effective news media among millions of Arabs; similar media outlets concerned with cultural, social, and educational events should be established.

The primary objective is to stimulate minds and encourage critical thinking for civilised debates and dialogues. Governments should control neither the news nor appointment of editors; quality and appropriateness should be controlled by the judgement of professionals and the wisdom of society in accordance with the rule of law.

We Arabs can accomplish the transition to the world of the 21st century, but the people and leaders must embark on a new course. Incremental changes — so-called gradual reforms — are inappropriate for a system that has been ineffective for decades. We should have confidence in ourselves and in global participation, and not blame others for current calamities or use religion for political gains. The responsibility of the individual for self and societal improvement is clearly stated in The Koran: "Indeed! God will not change the good condition of the people as long as they do not change their state of goodness themselves."

I appeal to the Arab people to participate in this process of historic change and not to be distracted by the ideologies of the past and conspiracy theories of the future. Being passive creates a state of apathy and legitimises the status quo. I call on intellectuals to focus on the greater good, not just for personal gain. Conscience and integrity are national responsibilities in this critical period of history. I urge the leaders of the Arab world to implement these historical changes and, in so doing, become makers of history. A genuine and peaceful transition to democracy is both legitimate and timely.

Before too long the oil will run out and human talent will migrate, but if we commit to "pillars of change", with jihad for modernity and enlightenment, we will realise our rightful place in the future.

The author is the only Arab Muslim to receive the Nobel Prize in Science, 1999.

Essays and Treatises*

* References to these Essays and Treatises are given in the Index at the end of the book.

Special Report: Egypt Today and Tomorrow

THE
CAIRO REVIEW
OF GLOBAL AFFAIRS

6 / Summer 2012   www.thecairoreview.com

EGYPT IN THE WORLD
By Nabil Fahmy

Brother President
by Shadi Hamid

The Second Egyptian Republic
by Tarek Osman

Islamism Now
by Ibrahim El-Houdaiby

Understanding SCAF
by Zeinab Abul-Magd

Cairo: A Memoir *by Éric Rouleau*

Tahrir Forum: *Ahdaf Soueif, Ahmed Zewail, Mohamed A. El-Erian, Jimmy Carter,*
Febe Armanios, Gro Brundtland, Rania Al Malky, Steven A. Cook, Khaled Fahmy,
Mona Prince, Michael Wahid Hanna, Malak Zaalouk, Shaden Khallaf, Mohamed Elshahed,
Ronnie Close, Tarek Osman, Rami G. Khouri, Nancy Okail, Daniel Williams

School of Global Affairs and Public Policy ⊛ The American University in Cairo

WE MUST DREAM

Replacing the Darkness of Ignorance with the Light of Knowledge

By Ahmed Zewail

When I came to the United States in 1969, I was not dreaming of a Nobel Prize, nor was I dreaming of acquiring a Bill Gates fortune. Armed with the excellent education I received in Egypt, I was simply on a quest for knowledge and a PhD degree from a reputable institution in the United States. At the time, my English was so poor that at restaurants I used to order "deserts" instead of desserts. America was a magnet for many members of my generation because of its leadership in science and technology and its unique democratic values. The historic landing of Neil Armstrong on the moon in 1969 was enough to demonstrate America's outlook regarding the frontiers of revolutionary knowledge.

△ Author Ahmed Zewail, Pasadena, June 11, 2011. *Christian Sabanpan for the Cairo* Review

I was aware of Edison's dictum, "Genius is one percent inspiration and 99 percent perspiration," and I took advantage of being in the right place at the right time—of being in America and at the California Institute of Technology, or Caltech. In fact, it was Caltech's ambiance and the country's system of support that made it possible for a young assistant professor to carry out, with his team, research that in only ten years' time would define a discipline that was recognized by the Nobel Prize in 1999.

People often ask me, "How does one get a Nobel Prize, and what is the secret to success?" I believe it was passion for science that supplied the energy, and it was optimism that made the almost-impossible possible. Success comes to the prepared mind. Success is not like rain that falls from the sky equally upon everyone: success is what you reap when you sow with passion and optimism.

Times have changed, the world is more complex, and the America of today is not the one I came to in the 1960s. We are now in the so-called "global age," threatened by chemical, biological, and nuclear disasters. The United States is facing real challenges: the rise of economic superpowers such as China and India, the conflicts and wars overseas, and—most importantly, in my view—the change in cultural, educational, and political values. Yes, there are challenges and changes. But, if I take today's Caltech graduates as an example, they can still make their own success in their own way. I would add that they are fortunate to have received an exceptional education in a twenty-first century, developed society. The education they received is unaffordable to at least 80 percent of the seven billion people on the planet who make less than ten dollars a day. And, just as importantly, America continues to provide them with opportunities that still cannot be found elsewhere in the world. And they are free to speak and worship as they please. And they can sleep at night without fear of the government or police. These fundamental values are embedded in the foundation of the American constitution, which is built on the pillars of life, liberty, and the pursuit of happiness.

Our world today is full of opportunity, and graduates have a unique role to play because of the special education they receive in the sciences and the rational thinking that this education has instilled. We should not listen to pessimists; but rather forge ahead to share the experience in whatever field we are passionate about, which could be business, government, law, art, or science.

I do not know the future of business or politics, but I know the future of science. This generation and the ones to come will continue to seek a basic understanding of nature and will make the many exciting discoveries that lie ahead—from deciphering and controlling the most fundamental constituents of matter to discoveries at our universe's boundaries and the unveiling of our origin and the miracle of life.

This generation will also explore other planets and possibly reach out to other galaxies. Part of Caltech's research is being done at NASA's Jet Propulsion Laboratory, which developed the Mars Science Lab rover named Curiosity. In August, Curiosity will perform the first-ever precision landing on Mars and help us assess whether the planet can harbor (microbial) life.

Beyond these and other intellectual achievements, there are direct benefits to people's liberty. From early in history, the quest for knowledge has been a driving force of revolutionary change, not only causing paradigm shifts in our understanding of the cosmos but also acting as an agent for the naissance and renaissance of human societies. The European Renaissance would have been impossible without the enlightenment regarding the significance of knowledge and rational thinking. I think too much credit is given to the impact of politics on the progress of society. Without science there is no development, and politicians would be unable to promise prosperity.

Just think: What would our world be like without electricity, penicillin, and the airplane? From the agricultural and industrial revolutions to today's genomics and IT revolutions, science has always been at center stage for societal advancement, and graduates must surely play a leading role in conquering the next frontiers of discovery, innovation, and progress.

Even in politics, technology is becoming the new weapon for transformative change in society. The youth of this generation are now harnessing information technology to do what those of my generation thought impossible. Elsewhere in the world oppression, occupation, and human suffering still exist but young people are rising up to acquire liberty from repressive regimes. The hope I witnessed—and am witnessing—in Egypt is a telling indication of a new role for science in democracy.

A people's revolution is sweeping the Middle East. I witnessed in real time the Egyptian uprising that began on January 25, 2011, and, remarkably led to the removal of Mr. Hosni Mubarak in only eighteen days. I saw university students in the hundreds of thousands, and then people in the millions, marching toward Tahrir Square in Cairo. The name of the square means "liberty" and that is precisely what the youth wanted from a thirty-year-old regime. They demonstrated peacefully, with impeccable organizational skills, and in unison. In my generation, we would probably have had to use stones, sticks, and guns in order to rise up; in this generation, they used Facebook, Twitter, and SMS. Without the development of the chip, wireless technology, and the Internet, this revolution may never have succeeded as a peaceful and civilized transformation.

Although the road ahead is bumpy, I am optimistic that, with investment in education and development through science, a democratic Egypt will emerge. Only a few months after the revolution began, Egypt announced the establishment of a new city for science and technology on three hundred acres of land, a national project that I have personally pushed for more than a decade.

When people in the Middle East ultimately gain their freedom, the world will be better off. Some scholars argue that the world is destined always to be embroiled in conflict and war. But this bleak picture is surely not the result of any natural phenomenon. We, the people, cause such conflicts and we, the people, can either kindle the fire or help extinguish it. The United States cannot change the culture of other people, and nations are responsible for their own plights, but it is the *kismet* of the United States to lead in the world by utilizing its most crucial force: the American value system of individual liberty, justice, and human rights. I believe that much can be achieved not by hegemony but by the strategic use of the real force of America— its soft power.

The soft power of science has the potential to reshape global diplomacy—and at significantly lower expense than that needed for use of the hard power of military involvement. I am hopeful that a new policy will be chartered for leadership in innovation. This policy should be inclusive of international science diplomacy for partnerships in development. Some may argue that it is naïve to think of applying such idealistic values in our imperfect world, but directing the influence of science diplomacy is in the best interest of the United States. Through the power of knowledge, we can efface ignorance and shape a future that binds cultures and civilizations.

In his 2009 Cairo speech, President Barack Obama articulated a new initiative for cooperation and partnership that emphasizes the role of science in diplomacy, particularly with Muslim-majority countries. Earlier, the president appointed me to his Council of Advisors on Science and Technology, and later I became the first U.S. Science Envoy to the Middle East. I embarked on a diplomatic mission that took me back to where I came from, but now with a different objective. From touring and seeing the state of science and education, not only in the region but also globally, I believe we will come to face serious consequences if we do not choose to act.

I recently read an important study that left me awestruck by the demographics of knowledge across our planet's population. In their book, *Educating All Children: A Global Agenda*, Joel Cohen and David Bloom argue that the aim of achieving primary and secondary schooling for all children is urgent and feasible, and yet more than 300 million children will still not be in school in the year 2015. Every effort should be made to change this state of affairs so that we may hope for a better future for our world.

However, education in the twenty-first century is far-reaching. It reaches beyond classical boundaries—not just across so-called interdisciplinary and multidisciplinary fields, but also between nations and maybe soon even across planets. Perhaps the best words to describe the value of education and knowledge are those written by Thomas Jefferson in 1782 [*Notes on the State of Virginia*, Q. XIV, 1782. ME 2:204]:

"The general objects [of a bill to diffuse knowledge more generally through the mass of the people] are to provide an education adapted to the years, to the capacity, and the condition of everyone, and directed to their freedom and happiness."

Remarkably, Jefferson, more than two centuries ago, saw the virtue of education on the individual and global level. And education remains a continuous process. Even university graduates are in the initial stage of a long voyage. During this journey, the wealth of knowledge should be wisely forged in place and time and opportunity. Having a dream and working hard to realize that dream, gives a meaning to life. Martin Luther King Jr. and other great men and women have realized these values. Without hard work, we are not entitled to a good life—and without compassion we will not attain the good life in a population dominated by have-nots. The investments of our families and our countries in education were made for good reason. We all need a good education to lead a fuller, richer life; our countries need educated citizens to build the future; and the world will be a better place when knowledge replaces ignorance.

My message to young people is simple: always be guided by the light of knowledge and wisdom to shape your future, the future of your country, and the future of the world.

This essay is adapted from Dr. Ahmed Zewail's commencement address at the California Institution of Technology on June 10, 2011. ■

The Arab Revolution

GRAHAM E. FULLER ▪ **AHMED ZEWAIL**

TARIQ RAMADAN ▪ **AYAAN HIRSI ALI**

BERNARD-HENRI LÉVY ▪ **OLIVIER ROY**

ABOLHASSAN BANI-SADR ▪ **MOHAMED DELKATESH**

RAMIN JAHANBEGLOO ▪ **EVGENY MOROZOV**

The Great Arab Revolt of 2011 has moved swiftly from the peaceful overthrow of autocrats in the nation-states of Tunisia and Egypt to brutal repression in the tribal societies of Libya, Syria, Bahrain and Yemen.

Meanwhile, the wired youth bulge of the Middle East that brought change is dissipating into an impotent diaspora while the organized interests of the old regimes and the once-suppressed Islamists charge ahead to power. This section examines the revolt, the reaction and the power struggles in its aftermath.

How to Jump-Start the Post-Revolutionary Era in Egypt

AHMED ZEWAIL *is the Linus Pauling Chair Professor of Chemistry and professor of physics at the California Institute of Technology. He also serves on President Barack Obama's Council of Advisors on Science and Technology and is the US Science Envoy to the Middle East. Zewail received the 1999 Nobel Prize in Chemistry for his pioneering work in the field of femtoscience.*

CAIRO — As I was leaving Cairo after Hosni Mubarak stepped down, I asked Esraa, a young woman who was one of the leaders of the revolution, "what was your objective?" She said, "taghier al nezam" — a change of the system. The Egyptians brought down the head of the system, but not yet the system itself. That is the challenge now. The Egyptian revolution, like that in Tunisia, represents a unique model for change in the Middle East. Because of history and traditions, these societies are not fragmented by tribal or sectarian conflicts. Despite differences of people's faith or even the occasional collisions between them, Egypt is united. It is not so much in the nationalist way as we have understood it in modern times, but in a civilizational way as "Umm al-Donia," meaning Egypt is "the mother of the cosmos."

In contrast, the second model for revolutions is that of Yemen, Libya and others in the making. In these cases, unfortunately, tribal and sectarian conflicts may lead to chaos and civil war, ultimately dragging the Middle East backward into conflict and fanaticism, not forward.

To avoid this outcome, the revolutions in Egypt and Tunisia cannot be allowed to fail. Egypt is the key. With 85 million people, it is the largest country in the Middle East and the heart of the Arab world. Making sure Egypt succeeds is essential for the stability of world's energy supply as well as for peace in the region. The West in particular must show its support for these peaceful uprisings for democracy, the reason that was claimed to have been behind the war in Iraq.

What can be done at this time? From my own involvement as an instigator of change for over two decades and as a negotiator with the youth and government leaders during the Egyptian revolution, I know what the youth want the most — a

Based on *The Financial Times* Op-Ed titled *"Fund Egypt's Future to Save the Arab Uprising."*

new future, different from what they had in the past, where education and development give them an opportunity in life. The so-called "children of Facebook" that
fomented the revolution know that Egypt was once ahead of South Korea in its
level of science and education. They know that in the 30 years Mubarak sat in his
palace and Egypt deteriorated, China has lifted hundreds of millions out of poverty,
sent astronauts into space, built glittering megacities and hi-speed trains and brought
its urban students up to world standards. They ask why Egypt can't do the same thing.

Of course, while this is the kind of change that ultimately matters, it will take
decades to get there. That is why the first priority now is offering a "candle of hope"
that can light the path forward.

I see a timely opportunity. While the Egyptians themselves must be the builders
of the new nation, they need some help in rebuilding sustainable institutions. The place
to start is with the first command of the Quran — READ. For Egypt today that means
a "Renaissance in Education and Development." An Egypt Fund for READ should be
established through a global partnership of private and government organizations.

This non-political fund should be directed by a board of trustees from renowned
Egyptians and world leaders in cooperation with the Egyptian government, and be
solely for the purpose of establishing a new education system and charting a capacity-
building strategy for economic development.

The aim of this fund would be complete reform of education from K-12 to the
university level to the establishment of research and development centers needed to
build up Egypt's infrastructure and the industrial complex.

Illiteracy is a major barrier for development and for fostering democracy. I would
suggest that READ focuses on managing and supporting, in a world class style, three
prototype and major projects — eradication of illiteracy, establishment of science
schools for the gifted, and building centers of excellence for R&D.

At the outset, such an effort would need to be seeded with $1 billion, along with
a $1 billion initial endowment to be drawn from private and government sources.
Over the coming years, further funds beyond this initial seeding could come from a
consortium of nations and be deployed through the World Bank, the Arab Bank and
the Islamic Development Fund.

Given that nearly half of Egypt's population is under 30 years of age, the return
on investment of knowledge-based economic development for a new Middle East
would dwarf the results of having spent trillions on the wars in Iraq and Afghanistan.

The psychological influence of launching such a fund at this moment cannot be
underestimated. Like Tunisia, Egypt today is in a "fluid state" in which a whole range
of interest groups, from the remnants of the old regime to political and religious

fanatics are contending for the direction of Egypt's future. By showing that effective institutions can be built successfully, the READ project would be a hopeful compass that can point Egypt in the right direction.

In Egypt, every family is suffering financially and emotionally from the deteriorated education system of Mubarak's regime. What families want most of all is to secure a good education for their children and work for a better future of the country. Tangible and immediate progress on this front will be critical to preventing Egypt from sliding backward to its old ways.

In reality, there is no other solution for Egypt's long-term future than better education and development. Egypt does not possess rich natural resources. Its agricultural area is relatively small — less than 10% of the total land. Its growth now relies on tourism, Suez Canal tariffs and foreign investment. Yet, according to the United Nations, Egypt's population will grow to reach 114 million before it stabilizes in the year 2065.

In history, recurrences do occur. If Egyptians are ready to take advantage of this unique moment, it can recover the greatness which made it a cradle of civilization and the intellectual and industrial center of the Arab world.

It is in the best interests of everyone — the Chinese, the Americans, the Europeans, the other Arab states — who want long term stability in the Middle East that the peaceful democratic revolutions in Egypt and elsewhere succeed. Revolutions seeking to create a new order are at most risk in their infancy. If we don't act now when the window is open, the consequence of not rapidly consolidating the benefits of change will haunt the Middle East for decades to come.

Time is of the essence!

A scientific revolution

The Arab Spring puts the Middle East in a position to become a scientific powerhouse, but it needs help from the west, says **Ahmed Zewail**

SCIENTIFIC research in the Arabian, Persian and Turkish Middle East lags behind that of the west. Of course, there are individual scientists who produce world-class research and there are institutions and nations which make significant contributions in certain fields. Publication and citation indicators show some encouraging trends. But naturally one asks: "Why have Arab, Persian and Turkish scientists as a group underperformed compared with their colleagues in the west or with those rising in the east?"

It is simplistic to say that there is a single cause, such as a (false) dichotomy between faith and reason. Muslims are no different from anyone else; there is no ethnic or geographic monopoly on intelligence. Muslims in Spain, north Africa and Arabia were at the peak of a sophisticated civilisation when Christian Europe was in the Dark Ages.

I think the answer lies in the recent history of the Arab, Persian and Turkish world. Consider what happened in the past century. First there was colonisation by western empires, which installed class and caste systems from outside. The result was huge populations of illiterate peasants. Illiteracy reached nearly 50 per cent, and among women it was as high as 80 per cent in many countries. When colonisation ended after the second world war, these countries looked to the superpowers for help, first west then east. And when the cold war ended, there was only one place left to look: up. That search for answers has been exploited by some to politicise religion.

It goes without saying that the developing world should help itself. The Middle East must not think itself incapable of competing with developed nations. But in addressing the gap, one must bear in mind a history that has resulted in large populations of frustrated people who lack real opportunity.

Many graduates in the Middle East are without jobs. What are their options? Their energy must not be allowed to be diverted into fanaticism and violence. In contrast to the silver wave faced by rest of the world, the Arab world is facing a youth wave. These young people can achieve great things in science if they are given the chance.

I see three essential ingredients for progress. First is the building of human resources by promoting literacy, ensuring participation of women in society and improving education. Second, there is a need to reform national constitutions to allow freedom of thought, minimise bureaucracy, reward merit, and create credible – and enforceable – legal codes.

The recent revolutions in Egypt, Tunisia and elsewhere show that these changes are possible. Over the past two decades I have been involved in promoting political and educational reforms, and I feel we now have an opportunity to make a real change.

Thirdly, the best way to regain self-confidence is to start centres of excellence in science and technology in each Muslim country to show that Muslims can compete in today's globalised economy and to instil in the youth the desire for learning. It is gratifying to see such centres being set up in Turkey, Malaysia, Indonesia, Qatar and elsewhere. In Egypt I am reviving the National Project for the Development of Science and Technology, which the Mubarak regime made every effort to derail despite the overwhelming support of the Egyptian people.

What can the developed world do? First and foremost it can partner with Middle East nations to improve their research capabilities. It can also offer aid, but only under certain circumstances. Aid packages are usually distributed among many projects with no follow-up, leading to diffusion of resources and a lack of impact. Better results can be achieved by directing a significant fraction of the assistance to programmes of excellence selected to build up both infrastructure and human resources.

Aid must also be depoliticised. The use of an aid programme to help totalitarian or undemocratic regimes is a big mistake. In the long run it is far better to be on the side of the people, not on the side of a dictator.

Such partnerships aimed at improving science and technology in the Arabian, Persian and Turkish Middle East are in the best interests of both the developed and the developing worlds, as knowledge-based societies are better equipped to be part of the world economy. They will also contribute to progress and enlightenment, and hence peaceful coexistence and a more civilised and truly global humanity. ■

Part of this article was adapted from the foreword to *Exploring The Changing Landscape Of Arabian, Persian And Turkish Research,* a Global Research Report published by Thomson Reuters

REFLECTIONS ON ARAB RENAISSANCE

A Call for Education Reform

By Ahmed Zewail

I recently read an important study that left me in awe of the knowledge demographics of our planet. In *Educating All Children: A Global Agenda*, Joel Cohen and David Bloom argue that while the aim of achieving primary and secondary schooling for all children is urgent and feasible, more than three hundred million children will not be in school in the year 2015. Empowering future generations with contemporary liberal arts education represents a significant challenge, even for highly developed nations. A year ago, President Barack Obama announced a major expansion of Educate to Innovate, a program to raise, in the coming decade, the level of American students in science, technology, engineering, and mathematics, or STEM, as the disciplines are collectively known. One of the new initiatives of the project is to spend more than $250 million of public and private funds to prepare ten thousand new teachers and retrain more than one hundred thousand others in the fields of math and science. As a presidential advisor, I know that the Obama administration has made all levels of education, an enterprise of nearly a trillion dollar budget, one of its top priorities.

For the Arab world, good education is critical for making our future. The failure of Arab education is one of the underlying causes of the youth discontent we are witnessing throughout the region. The serious cultural, economic, and political consequences have become obvious. The children of Facebook have ignited an intifada to plant democracy in Egypt, Tunisia, Bahrain, Yemen, Libya and other Arab countries. Only when we diagnose the symptoms can we cure the disease, and it is education that is at the core of any recovery—an Arab renaissance. It is imperative and a matter of urgency that education be elevated to a much higher national priority throughout the Middle East. Anyone who examines the progress achieved in Europe, the United States, or in more recently developed nations

▷ Physics class at The German University in Cairo, March 1, 2006. *Gary Knight/VII/Corbis*

in Asia or Latin America, can understand the direct correlation between good education and the development of societies. Indeed, Egypt is a living testimony to the link between the power of knowledge and the impact of its ancient civilization—and that later one established on the shores of the Mediterranean some two thousand years ago, in Alexandria, whose library and museum constituted a beacon of knowledge.

The concept of civilization itself is based upon knowledge. On the banks of the river Nile, the ancient Egyptians introduced new concepts in fields such as architecture, medicine, astronomy, and chemistry that influenced the Pharaonic as well as later civilizations. In my own field, some six thousand years ago, they were the first to ingeniously measure time and create the solar calendar. Recently, French scientists reported yet another astounding achievement, namely that the eye cosmetics used in the time of Nefertiti contained a man-made lead compound that helped treat or prevent eye disease. The modern West continues to explore the progress attained by this ancient civilization.

Arab civilization, a millennium ago, recorded outstanding achievements as well. Last year, I published with Sir John Thomas of the University of Cambridge a book about four-dimensional microscopy imaging. In this book and elsewhere I point out that major contributions to the science of imaging and vision were made by the Muslim thinker Ibn al-Haytham (known in the West as Alhazen) who lived ca. 965–1040 in Iraq and Egypt. He developed concepts in optics, later used by Descartes, Newton, Da Vinci, and our modern photographers, that explain how the retina works to receive an image from reflected light. His experiment, called *al-hugra al-muzlima*, "the dark room," later known as camera obscura, demonstrated how external light passing through a pinhole in a box forms an upside-down image on an interior surface. The making of useful knowledge by Ibn al-Haytham, Ibn Rushd (the polymath philosopher known in the West as Averroës), Ibn Sina (the foremost physician of his time, known as Avicenna), al-Khawarizmi (whose Latinized name, Algoritmi, inspired the terms algorism and algorithm), and other renaissance men, forged centers of enlightenment in capitals such as Baghdad, Cairo, and Cordoba.

Modern Egypt, too, was recognized for a renaissance in educational, cultural, and industrial fields, in part due to the progress made possible by the visionary leadership of Mohammed Ali. He transformed Egypt into a regional industrial and military power through reorganization and reforms in the society and by means of educational and cultural missions from Egypt to Europe. Rifa'a al-Tahtawi and his followers were among the pioneers in bringing about a renaissance in education with the aid of knowledge translated from French into Arabic. In the years to follow, Egypt became a powerhouse in literature, arts, science, and culture. Personalities in all fields emerged and influenced Arab society at large. We still live on the echoes of their contributions, from the writings of Taha Hussein, Naguib Mahfouz, and Ali Moustafa Mosharafa, to the songs and films of Umm Kulthum, Abdel Wahab, and Faten Hamama. In the Middle East, only a century ago, Egypt pioneered democratic governance and established institutions in different sectors of higher education and scientific research (Cairo University), banking (Bank al-Ahly), mass media (Al-Ahram), and industries such as textile and the motion picture. With this advanced status, Egypt attracted and educated future generations of Arabs, including, among many others, the renowned Palestinian Edward W. Said, Michael Atiyah of Lebanon, King Hussein of Jordan, and my own father-in-law Chaker El-Faham of Syria. As recently as the 1960s, after the 1952 revolution, my generation benefited from a fine education system amid a rich cultural life and national dreams of colossal projects, such as the High Dam, space aviation, and nuclear energy.

A Bleak Situation

With these past achievements in mind, and knowing that human resources are still available, one has to ask what happened to education and scientific research in the modern Arab world? Today, the contribution of the Arab world to international scientific research is negligible. In both research investment (R&D spend compared to GDP) and capacity (researcher numbers compared to population), as reflected in research publication output between 2000 and 2009, the predominance of Turkey and Iran is evident; in 2009 Turkey produced twenty-two thousand publications as compared to five thousand in 2000, and Iran produced fifteen thousand papers in 2009, in vast contrast to thirteen hundred in 2000. During the same period Egypt and Saudi Arabia had a flat trajectory at nearly two thousand, according to a very recent global research report by Thomson Reuters. No Arab university regularly ranks among the world's five hundred best institutions, though Alexandria University made it in one ranking in 2010. Ibrahim El-Moallem, vice president of the International Publishers Association, estimates that book sales in the Arab world are 1 percent of the world market (Egypt accounts for 0.4 percent), and that books of basic sciences and arts are only 3 to 5 percent of total sales. Egypt's national income from research and development is minor. The GDP depends on revenues from traffic in the Suez Canal, tourism, and natural resources such as gas and oil, all a "gift of the Nile," as Herodotus called the land of Egypt. By contrast, Israel, with less than a tenth of Egypt's eighty-million population, acquires more than 90 percent of its GDP primarily from industry and services, with technology being the back bone of the economy.

Current conditions do not encourage optimism if we consider government expenditures. On basic education, according to a report by the renowned medical doctor Mohamed Ghoneim, based on United Nations Development Programme data, the

Egyptian government spends annually twenty-four billion Egyptian pounds (less than $5 billion and around 2.4 percent of the national income) which amounts to about $250 per student per year. Israel, to make another comparison, spends at least $1,500 per student per year. At university level, the Egyptian government spends less than $500 per year on each of its 2.2 million students. In contrast, Egyptians are paying $15,000 and $10,000, respectively, for private education at the American University in Cairo or the Nile University, for example. Such is the bleak situation in public education that causes Egyptian families to pay between ten and fifteen billion Egyptian pounds per year for private tutoring in the hopes of giving their children a greater chance to learn and succeed.

The infrastructure of most of the Egyptian schools is far behind a country like Finland or South Korea. The number of students in public-school classes reaches sixty and more, making it impossible for a teacher to interact well with pupils. University lecture halls are packed with hundreds of students. Teaching also relies heavily on indoctrination, failing to take advantage of pedagogical methodologies that have advanced throughout the modern world. Many of the topics in the curriculum are not suitable, especially when we consider that these students have to compete in the information and space age.

In Egypt, teachers have an unfavorable financial position and a lower-than-ever social status. Even with a recent salary increase, many find themselves forced to offer private lessons on the side as a way of supplementing their incomes. What is needed are teachers, or *mu'allims*, who can cognitively attract students to new ideas and knowledge, not simply tutors who prepare them for rote memorization and passing exams.

Society and Media

Conditions in Arab society at large have not been very conducive to advancement in education. More than 35 percent of the Arab population is under the age of fifteen. This represents a potential boom in human resources, yet it is not effectively utilized. The unemployment rate in the region ranges from 10 to 20 percent. Egyptian families face a greater burden in raising children today compared to fifty years ago. In an estimated 30 percent of the families, women are the breadwinners due to issues like divorce.

The Arab media are not equipped to perform their responsibility to illuminate educational and scientific matters, despite the increasing number of newspapers, magazines, and over five hundred television satellite channels (most of them concerned with music video clips and soap operas). A few years ago I happened to arrive in Egypt right after the landing of the NASA rover, Opportunity, on Mars. The world seemed transfixed by this remarkable event, but in Egypt I could only find a small story about it in one of the newspapers. It is generally acknowledged that the search for truth and in-depth analysis are uncommon features in the programming of the influential broadcast media, which consume about one-third of the day in the life of the average Egyptian. In a recent TV program watched by millions, the anchorman said we must find out "who changed the weather" (*min illi ghayyar al-gaww*) to make it so bad, adding in all seriousness "We must deal immediately with the culprit or the country involved!"

The culture is under pressure. One of my concerns is that we are creating sub-cultures within the Arab culture. In private schools, which are now becoming a major force in Arab education, the language used and culture practiced are often those of other nations. Of course it is proper to teach foreign languages, yet without proper education in the Arabic language and traditions, the country risks class/language fragmentation in society. I also see blind imitation as another threat to the culture of the Arab world. If MTV in America (which incidentally does not reflect the rich American culture) airs certain programming, it does not mean that Arab society is not modern unless its channels offer the same thing. The quest for modernization must respect cultural identity.

In the realm of religious values, the vital role of true religion has been replaced by ideological stubbornness and, I may say, ideological terrorism by some. I believe that the use of politics in religion and religion in politics is creating confusion and conflict in society. The spread of rulings or pronouncements from non-qualified proselytizers has diverted society's energy into superficial issues and led to calcification of the mind. And evasion of the rule of law has had many consequences. The most threatening of all is 'societal fragmentation,' which has led to cracks in national unity through bigotry and violence, as we saw in Nag Hammadi and Alexandria in Egypt. Superficial media rhetoric does not solve problems and people must use reason to reach their goals for national coherence and civil society governance. One essential change is good education.

A Paradigm Shift

It is clear that the education system is in need of major reform. In any nation, schooling can generally be categorized into primary and secondary education, higher education, and research and development, normally involving Master's and Ph.D. degree studies.

Primary Education

Basic education is a human right, especially in the knowledge society of today. It must be a priority for governments to embark on a major national project to eliminate illiteracy. I find it hard to believe or understand, in an age seeking ever greater scientific knowledge, that from 25 to 50 percent of the Arab population of some three hundred and fifty million remains illiterate. How can this population at large deal with the modern world of services through the internet, or be effective in the knowledge-based work force? How can illiterate parents prepare the children of tomorrow?

Arabs need to dramatically increase spending on education. In most of the Arab countries, the current low percentage of national wealth spent on education will never improve the schools, curricula, and teachers to produce the kinds of students that are equipped to handle national and international demands. More resources are needed, and they should have priority over less essential projects like mega-resorts for luxury living. The allocation of new resources must be directed with care to enhance quality over quantity.

The status of teachers must change through a merit-based system of evaluation and appreciation. That should eliminate the present parallel education system of private tutors, upon whom Egyptian middle-class families spend a disproportionate amount of their income to enhance an education that is supposedly free. The involvement of the private sector should take on a new structure, one that creates a partnership between schools and families and involves them in the education process; no single person, not even a minister, knows the answers to questions pertinent to the most effective mechanisms in education.

Higher Education

Higher education with hundreds of thousands of university students and low-level resources has proved to be ineffective. It is time to restructure the current organization, which was effective half a century ago, and to create a modern multi-tier system of public and private higher education. For public education, it would be reasonable to establish three or four tiers of universities with the layers being defined according to students' ability and social situation. On the other hand, private universities should not be for business; in fact, they should represent the pinnacle of education, research, and development as nonprofit organizations.

In California, for example, the state supports the University of California and California State University systems, community colleges, and others such as adult education schools. Each of them has its own goals and mission; the aim of educating a student going to Berkeley is different from that at Cal State, Los Angeles, or Pasadena City College. In the end, the students will get educations that serve them and the society. But depending on ability, status, and presence on campus (full/part time), the student and the state are in a position to tailor needs and effectiveness. Parents who can afford their children's university fees should be required to make financial contributions, just as they do when their children are enrolled in private universities at home or abroad, but those students who do not have the means should acquire appropriate education either through merit-based scholarships or from a loan-granting banking system.

Also in California, again by way of example, there exist some of the best private universities in the world, including Caltech and Stanford. Besides the excellent education they offer, and the enabling of the thinkers behind Intel, Google, and other mega-companies, they represent powerhouses for global research and development. Both Caltech and Stanford are nonprofit private universities. Their support comes from endowment, philanthropy, and relatively high tuition. For research, both public and private institutions receive funding mostly from federal government agencies and the private sector. The National Science Foundation is a key source of funding, and Arab countries should establish similar entities for funding independent and creative research. I also suggest defense ministries develop funding programs, as a percentage of their total budget, for national support of R&D even if the research is only remotely related to the defense issues of today.

In the region, Israel and Turkey have succeeded in establishing advanced private institutions. In Turkey, Bilkent University is a leading research institution, whose endowment is sustained by income from major international projects such as the construction of world-class airports (for example in Cairo and Doha) and the supply of high-tech products to various industries. Such a concept was proposed as a national project in Egypt more than ten years ago; more details can be found in my book, *'Asr al-'ilm*. With independent institutions, Israel has become a high-tech superpower over the past two decades. Scaling to its population, it leads the world in number of start-ups and size of venture-capital industry. The Israeli government has now identified new frontiers of focus, including areas of potential growth, in alternative energy, water management, agricultural innovations, and of course in the military industry.

Teaching methodology in higher education is in need of revamping. Understanding of and respect for facts is essential for the scientific method, which is not only important for education itself but also for integrating rational thinking into the fabric of the culture.

In countries with overpopulated universities, students are not provided with "hands on" opportunities for involvement in learning. When I was a student at Alexandria University, I was in a special class of seven students and had direct access to microscopes, experiments, and the like. Today, with the large number of students involved, carefully instituted and interactive teaching methods can enable students to perform experiments through virtual reality technology. Such methods, which I have observed in Turkey and Malaysia, represent a totally different and better way for cognitive involvement than teaching students science largely through memorization. Faculty must be equipped with the latest techniques, through retraining and periodic updating, and perhaps through sabbatical leave. Such faculty are to be evaluated and academically and financially rewarded according to performance.

To increase the level of skill nationwide, technical and vocational education must be enhanced and respected. This type of education empowers the society with know-how and improves the infrastructure. It is also time to stop admitting more students than universities have the capacity to teach, and to end the exaggeration of scores in obtaining degrees. In the 1960s, it was a great achievement for me to score over 90 percent in my bachelor's degree. Now we hear of 110 percent scores! Scientifically, 100 percent is the maximum meaningful score. I can understand so-called bonuses, but these should not be confused with actual grades.

Finally, the reform of higher education should begin with the restoring of the prestige of the university and faculty and by raising the admissions standards. I recall how emotional I felt on my arrival at the campus of the Faculty of Science, because of the high standing, the *hayba*, the university and faculty had in our society at that time. The campus should be the home of knowledge and culture, not a space for political and religious conflicts. It need not be policed for security. In all universities I am familiar with, students can have intellectual discourses on all subjects and organize all kinds of events under the supervision of their mentors and subject to university by-laws, and, if extra precautions are needed, these take the form of campus security staff, not a police force.

Research and Development Centers

The third and final level of the education pyramid is scientific research. First, we should clarify a few misconceptions. No nation can establish a viable R&D program without commitment from its government in the form of long-term investment. In fact, without such backing I cannot see why the public in general and the private sector in particular should contribute to funding R&D. Second, developing nations can achieve progress on the international level in a relatively short time, contrary to the belief that such progress requires one or two generations. The proof came recently from a number of countries including China, India, South Korea, Malaysia, and now Turkey and Iran. Third, R&D is not a luxury, as some believe, reserved for rich or developed countries. This too is proved false, since developing, and in many cases poor, countries have crossed the chasm that divides them from developed nations by investing *more* generously in R&D. Fourth, there is a fundamental role for basic research in development, and this cannot be strong without a science base that integrates expertise in STEM.

Curiosity-driven basic research requires that creative scientists work in an environment that encourages interactions between researchers and collaborations across different fields. But such attributes cannot and should not be orchestrated by structured and weighty management, as creative minds and bureaucracies are incompatible. Large buildings alone will not produce much without the right people. To distract faculty members with excessive regulations, research-for-promotion incentives, or to involve them as political tools, is the beginning of the end. Without resources little can be achieved, no matter how creative the mind. Countries and institutions that provide the proper infrastructure and the funding for new ideas will be the home of new discoveries and the source of innovations.

The quest for new knowledge drives innovation, and without centers of excellence, young students and scientists will not be attracted to the profession of R&D. My own attraction to science was enhanced by the sheer joy of discovery, which began in Egypt with a school-age curiosity about why wood burns. When I look back and ask why Caltech, or the California Institute of Technology, where I have spent the last thirty years of my career, has garnered thirty-five Nobel prizes, it is because Caltech as an institution believes in such values. Its unique culture makes scholars enjoy the quest for the unknown, free from bureaucratic regulations and political hierarchy.

It is time for Egypt and other Arab countries to have such a vision and for governments to be determined to allocate substantial resources to establish R&D centers of excellence. These centers should be rising with at least equal priority to the media megacities recently brought into existence in several Arab capitals. As importantly, the centers must be granted independence in order to formulate their own academic and administrative structures. Among other things, establishing a merit system for scholars and elevating the prestige of the chair professor are important essentials. For young researchers, the government should increase the number of scientific missions it sends abroad. In turn, the research environment should be attractive to the scholars when they return to their homeland with the benefits of the knowledge they have acquired. Their missions abroad are a waste of money if they are forced to struggle with bureaucratic obstacles back home.

The Knowledge Chasm

In the Arab world, we face the daunting challenge of reducing the knowledge gap with the rest of the world in many fields of endeavor. When I was a graduate student, America, after the launch of the Russian sputnik, had a vision to conquer outer space, and in 1969 Neil Armstrong walked on the moon. Today, America is sending space cars and possibly an astronaut to Mars. Scientists are searching for life on other planets with the potential of finding new resources or, in the case of some governments, with the desire to control the Earth from outer space.

In the field of medicine, discoveries are made at the level of genes and cells, opening up new avenues for drug design and cures for disease. Today, scientists can take adult cells from skin and convert them into stem cells that can be used to develop new tissues for the heart, eyes, and other parts of our bodies. Such discoveries were unimaginable twenty years ago.

In the microscopic world of nanotechnology, for the first time in human history, man can visualize objects in four dimensions, functioning on a time scale of a millionth of a billionth of a second, and in their 3-D space with a better than one-billionth-of-a-meter precision. This methodology, which my group and I developed recently at Caltech, has the potential of uncovering new phenomena in materials science and in biological/medical sciences.

These are just a few examples that demonstrate the chasm between the haves and the have-nots of knowledge. It would be naïve to think that the quest for useful knowledge is a luxury for society. Exploring for its own sake enables humans to discover the "unknown unknowns," and not simply to polish our knowledge of the "known unknowns."

Epilogue

Civilizations rise through the power of knowledge. They fall when such power fades. Arab and Muslim civilizations reached their zenith when their leaders believed in the value of making new knowledge and in ensuring human rights and liberty of mind, the necessary tools for progress. Preserving knowledge is easy. Transferring knowledge is also easy. But making new knowledge is neither easy nor profitable—in the short term. It has, however, proved to be hugely profitable in the long run. Think of the impact of only two curiosity-driven discoveries, the laser and the transistor, which have now transformed world markets and human services. And there is more: knowledge is a force that enriches the culture of any society with reason and basic truth and enlightens people against bigotry and radicalism.

In the modern era, the Arab people have not been makers of useful knowledge, and they are facing major challenges. Some of these are crippling their influence and participation in the world market and others are threatening the foundation of their own culture. At the core of the problem is the challenge of a deteriorating education system. Without education there is no progress and progress does not lie in the ability to consume and acquire goods from abroad. Arabs are in need of a renaissance that is built on a modern education and a science base with its triad of basic research, technology transfer, and societal involvement.

The impossible is possible. I certainly have confidence that Egypt can succeed despite all the complexities and problems in education and scientific research, and in governance. Turkey and Iran are emerging as a real force in the Middle East, and without Egypt's force the Arabs will not be in the sphere of influence. Some Arab and Muslim countries, including Qatar, Turkey, and Malaysia, have already made progress in the field of education, transcending the stereotypes currently associated with Muslim culture and religion. But no country can cut corners to development. Important changes and progress in the Arab world will only occur if there is a political vision and will from the highest levels of the state. In the past, and we in the Arab world have all experienced this, it was possible to enclose a whole country. Today, that is impossible. The poor and the rich alike have small satellite dishes on their roofs, which show them how things are around the world. The sky and the whole world are open to our children. The political upheaval in the Arab world is telling of the need for a speedy change.

As we work toward a better future, there should be no conflict between science and religion. Education, which etymologically derives from the Latin "to bring out" potential, is any act or experience that has a formative effect on the mind, character, or physical ability of an individual. Education eradicates ideological stubbornness. In Arabic the word *ta'lim* captures its essence; it is from *'ilm*, or knowledge. Education is the process by which society deliberately transmits its accumulated knowledge, skills, and values from one generation to another. It is therefore in the fabric of civilization. We should make use of the minds humans are uniquely blessed with, and accept faith and reason as the bases for human life. Egypt utilized both in building its ancient civilization. It must do so again.

Taha Hussein, one of our great writers, said decades ago, *"al-'ilm ka-l-maa wa-l-hawa'."* ("Education is a necessity, like the water we drink and the air we breathe.") Without *'ilm*, there is no life. Arabs have an opportunity to regain their place in history. The reasons for backwardness are several. Colonization and occupation have certainly been part of the problem. The self interest of superpowers, and even uninformed prejudices against cultures or ethnic groups, have always been in the fabric of politics. But we cannot live in the past or in the present with conspiracy theories. We must first solve our in-house problems in order to light the future. With "Liberty, Knowledge, and Faith" (the motto of the Ahmed Zewail Foundation for Knowledge and Development in Cairo), I believe that Arab renaissance will launch the dawning of a new age. The choice is ours: either to become three hundred and fifty million people of the cave, *ahl al-kahf*, or three hundred and fifty million people of the cosmos, *ahl al-kawn*.

This article is based on addresses given by the author at the Dubai Press Club and at the Cairo Opera House in 2010.

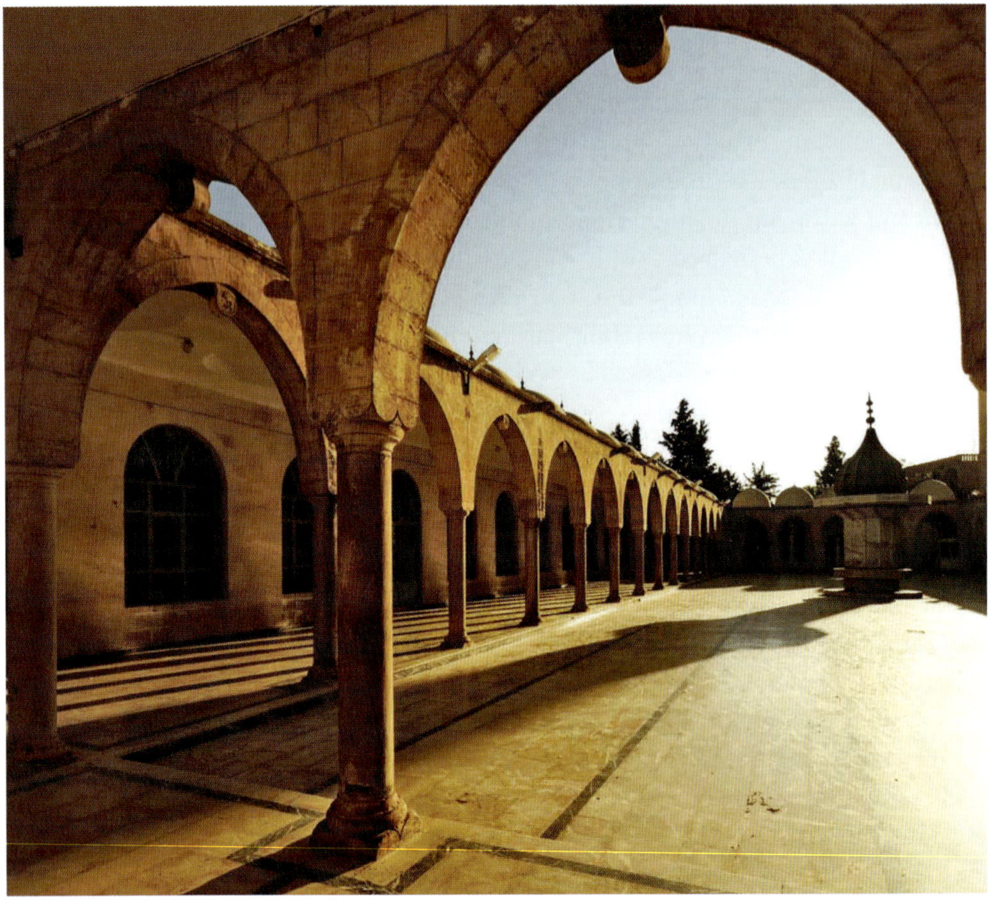

GLOBAL RESEARCH REPORT
MIDDLE EAST

EXPLORING THE CHANGING LANDSCAPE OF
ARABIAN, PERSIAN AND TURKISH RESEARCH

FEBRUARY 2011

JONATHAN ADAMS
CHRISTOPHER KING
DAVID PENDLEBURY
DANIEL HOOK
JAMES WILSDON

FOREWORD BY AHMED ZEWAIL

EVIDENCE

As this Global Research Report from Thomson Reuters summarizes, scientific research in the Arabian, Persian and Turkish Middle East lags that of the West. Of course, there are individual scientists from the region who produce world-class research and there are institutions and nations in the Middle East which make significant contributions in certain fields. In fact, the publication and citation indicators described herein show some encouraging trends for the region during the last decade. But naturally one asks, "Why have Arab, Persian and Turkish scientists as a group underperformed compared to their scientific colleagues in the West?"

It is too simplistic to say that there is a singular cause, such as a (false) distinction between faith and reason. From a genetic point of view, Muslims are no different from anyone else. There is surely no ethnic or geographic monopoly on intelligence. Historically, Muslims in Spain, North Africa, and Arabia were at the peak of their civilization when Christian Europe was in the Dark Ages. But what is more important is the modern history of what happened in the Arab, Persian and Turkish world. What happened in the last century? First of all, there was colonization, which installed a class and caste system of the governing elite from or allied with the outside; for example: the British Empire. The result: a huge population of illiterate peasants. Illiteracy reached 50%. For women it was as high as 80%. With this level of illiteracy, when colonization ended after World War II, what did these countries do? They looked first West, then East, to the superpowers for help. And when there was no progress in terms of economic development, after the Cold War, there was only one other place left to look: up. And that search for answers has been exploited by some to politicize faith and religion.

It goes without saying that the developing world should help itself. The Middle East must not think it is incapable of competing with developed nations in science and technology. I have written about this subject and have called for a new education jihad for acquiring useful knowledge. But in addressing the gap in performance, one cannot fail to bear in mind a history that has resulted in large populations of frustrated people who lack real opportunity. Many college graduates today in the Middle East are without jobs. What are their options? All their energy

must not be allowed to be diverted into fanaticism and violence. Unlike the rest of the world – facing a silver wave – the Arab world is facing a youth wave. These young people can achieve great things in science, if they are given a chance.

I see three essential ingredients for progress. First is the building of human resources by eliminating illiteracy, ensuring active participation of women in society, and improving education. Second, there is a need to reform national constitutions to allow freedom of thought, the minimizing of bureaucracy, the development of merit based systems, and the creation of a credible – and enforceable – legal code. Finally, the best way to regain self-confidence is to start centers of excellence in science and technology in each Muslim country to show it can be done, to show that Muslims can indeed compete in today's globalized economy and to instill in the youth the desire for learning. Several new centers of advanced education and research and development in the region represent such efforts. I am encouraging the development of a University of Science and Technology in Egypt with the same goal.

And what can the developed world do? It can partner with Middle Eastern nations to improve their research capabilities. In terms of aid, developed countries should focus assistance. Usually an aid package is distributed among many projects, with a lack of follow-up leading to diffusion of resources so that the aid does not really have significant impact. Real focus can be achieved by establishing what I call partnership-guided aid, with a significant fraction of assistance being directed to programs of excellence using criteria set by the developed nations. There must also be a minimization of politics in aid. The use of an aid program to help specific regimes is a big mistake, in my view. Developed nations can either give money as charity or – and this is better – they can become partners, providing expertise and a follow-up plan.

Such partnerships aimed at improving science and technology in the Arabian, Persian and Turkish Middle East are in the best interests of both the developed and developing worlds since they will significantly contribute to peaceful coexistence and a more civilized and truly global humanity.

Foreword by Ahmed Zewail

Further reading from the work of A Zewail:

"Curiouser and Curiouser: Managing Discovery Making," *Nature*, 468: 347, November 17, 2010 (http://www.nature.com/news/2010/101117/full/468347a.html)

"The US Needs a New Soft Era," *The Guardian*, July 11, 2010 (http://www.guardian.co.uk/commentisfree/cifamerica/2010/jul/11/soft-power-us-middle-east)

"Science as a Shaper of Global Diplomacy," *Los Angeles Times*, June 27, 2010 (http://www.zewail.caltech.edu/LA_Times.pdf)

"We Arabs Must Wage a New Form of Jihad," *The Independent*, August 24, 2006 (http://www.independent.co.uk/opinion/commentators/ahmed-zewail-we-arabs-must-wage-a-new-form-of-jihad-413101.html)

Voyage Through Time: Walks of Life to the Nobel Prize, World Scientific, 2002.

Mediterranean Scientopolitics

ON THIS YEAR'S BASTILLE DAY IN JULY, THE PRESIDENT OF FRANCE, NICOLAS SARKOZY, INAU-gurated a new initiative for uniting the Mediterranean South with Europe in general, and France in particular. The aim of the Mediterranean Union (MU), an analogue of the post–Cold War European Union (EU), is to "lay the foundations of a political, economic and cultural union founded on the principles of strict equality." Comprising 27 EU members and states from the Middle East, North Africa, and the Balkans, the MU would in principle unite close to 800 million people. In June, a meeting was held at the Institut de France with representation from many academies, scientists, and politicians to discuss possible cooperative programs. The goals expressed at the meeting are admirable; however, the MU's motives need to be clearly defined, as the issues for the MU are very different from those for the EU. Most important, thus far missing in the fabric of the former is an explicit role for education and science.

The Mediterranean people have a rich history encompassing cradles of civilization ranging from Egypt and Greece to the Roman Empire. As the word implies in Latin, the Mediterranean was considered the "Middle Earth," but at present the disparity between North and South is alarming. The difference in gross domestic product between the two is staggering, and illiteracy, deterioration in education, and the unfavorable state of governance in the South have put many there at a disadvantage. Despite these challenges, the MU could redefine the state of North-South cooperation by providing new opportunities for progress—but only if differences and concerns are openly addressed.

The integration of Eastern and Western Europe is to some extent easier than that of the North and South Mediterranean because religions and cultures are more diverse in the Mediterranean Basin. For the initiative to succeed, the leaders in MU nations must promote economic and political strategies that respect these differences. The benefits of free trade and liberty, coupled with dialogues of cultures through scholarly discourse that promote mutual acceptance, will undoubtedly lead to stronger bonding among and security for the nations in the region. But if the MU is directed by political agendas, such as distancing Turkey from the EU or isolating particular countries in the EU or Arab League, it will ultimately unravel and become a medium for slogans and the polarization of nations. The main political objective of the MU should instead be the promotion of human rights and liberty, and the solution of chronic problems such as the Israeli-Palestinian conflict.

The driving force essential to any progress is education and the ensuing scientific and technological development. General education will not only improve the well-being of society on all levels, but will also create an atmosphere of enlightenment that resists dogmatic and radical practices. Modern education will provide new skills and economies and the means for positive participation in the world market. And building a strong science base for research and development in the Mediterranean Basin, especially in the South, will have real impact with mutual benefits, allowing scientists and the rest of civil society to work together to alleviate many problems of significance to the region such as illegal immigration, illiteracy, food shortages, energy demands, water resources, climate change, infectious diseases, and the dearth of democratic governance. Moreover, creating such a base through sustainable cooperative programs with the North will limit brain drain and channel the energy of youth into a knowledge-based world economy.

The new MU initiative could turn into a historic milestone, building on the 1995 Euro-Mediterranean Partnership (the Barcelona Process), provided that there is a genuine desire for North-South support and partnership. Building education and the science base, bridging cultures through strong collaborative programs, and boosting economic and political benefits are the triad on which the MU should stand. These objectives will not see the light of day if the purpose of the MU is mainly political—rather, the focus should be "scientopolitical," a phrase coined here to emphasize the importance of education and science to the advancement of political and human affairs.

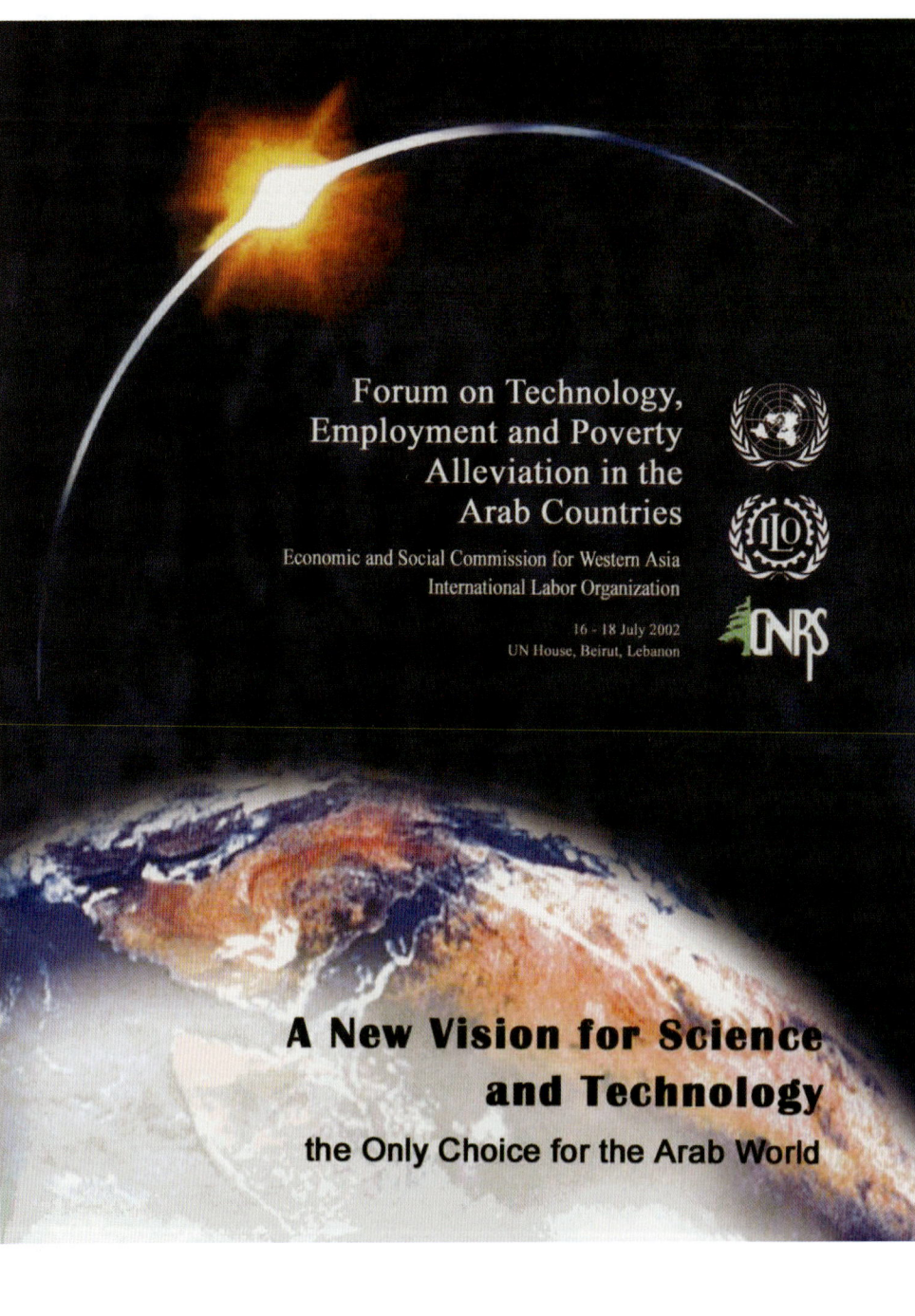

Forum on Technology,
Employment and Poverty
Alleviation in the
Arab Countries

Economic and Social Commission for Western Asia
International Labor Organization

16 - 18 July 2002
UN House, Beirut, Lebanon

**A New Vision for Science
and Technology**
the Only Choice for the Arab World

A new Vision
for Science
and Technology
The Only Choice
for the Arab World

Keynote Speech
at the United Nations Forum
on Technology, Employment
and Poverty Alleviation
in the Arab Countries,

Beirut, July 16, 2002

By:
Ahmed Zewail
Nobel Laureate

California Institute
of Technology
Pasadena, USA

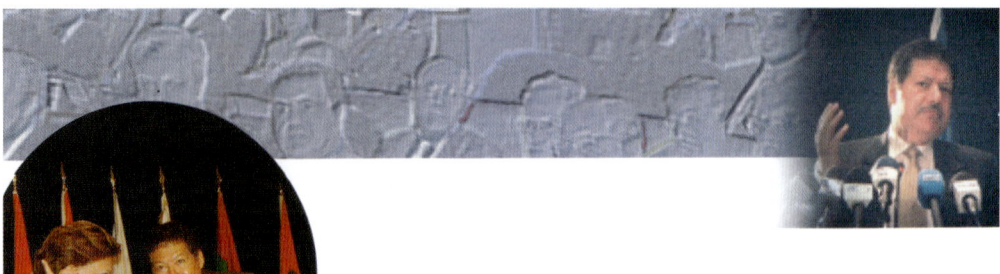

Advances in science and technology and the incessant march towards a global, knowledge-based economy pose both numerous opportunities and major challenges for developing countries, including countries in the Arab region. There is a need to identify and penetrate the new niche markets that are being created in a number of scientific fields, particularly the information and communications technologies (ICTs), biotechnology, gen   etic engineering and innovations in the science of materials. Additionally, technological developments promote corporate competitiveness and productivity, which creates employment and alleviates poverty. Moreover, when diligently linked to socio-economic and environmental issues, new technologies offer prospects for decent work and improved working conditions in both the developed and developing countries.

In light of these technological developments, the Economic and Social Commission for Western Asia (ESCWA) and the International Labour Organization (ILO) co-organized the Forum on Technology, Employment and Poverty Alleviation in the Arab Countries (Beirut, 16-17 July 2002). The overall objective of the Forum was to review and debate policies, strategies and initiatives in new technologies aimed at creating employment, alleviating poverty and improving work conditions in the Arab countries.

With a view to harnessing national capacities in science, technology and innovation (STI) in the Arab region, the participants in the Forum agreed on the need to create an enabling environment for sustained socio-economic development, which includes appropriate labour policies aimed at generating better employment and economic opportunities in the region, particularly for youth, women and small and medium-sized enterprises (SMEs). Moreover, international and regional developments in STI have created a sense of urgency. A new vision is now needed to craft national policies that can capture the promises of new technologies and that can ultimately lead to a renaissance.

The Nobel Laureate, Professor Ahmed Zewail, gave the keynote speech at the opening session of the Forum. He emphasized the need for this new vision in STI, which must be ambitious in leading to a veritable renaissance in the Arab region. The need to develop local technology capacities is crucial given that technology cannot simply be bought or imported and adopted. Moreover, in order to encourage innovation in the Arab region and to create an enabling environment, impediments to freedom of thought and expression must be removed. In practice, this requires the removal of such obstacles that exist in legislative, regulatory and bureaucratic systems. Professor Zewail concluded that the Arab countries must invest in developing human resources and create an innovation-friendly culture to propel science and technology forward in order to foster economic growth and improve living standards.

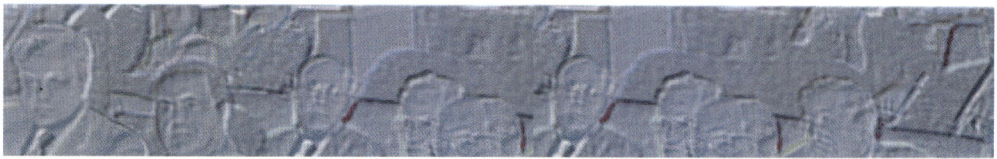

Reinforcing his speech, the recommendations from the Forum highlighted the need to establish an enabling environment that allows and rewards innovation in all walks of life, and to set up an advisory board and task forces to assist in policymaking. Within the context of this new vision in STI, the Arab countries need to implement a new system of laws and development policies that promote STI capacity building in order to meet the challenges and opportunities of the global, knowledge-based economy. Furthermore, there was a general consensus among the participants in the Forum that decent work and better living conditions are key objectives for Arab societies, particularly with due reference to such marginalized groups as women, youth, the disabled, the elderly and the poor in rural and remote areas. Within the framework of a competitive economy, all development policies in the social, economic and STI spheres must therefore converge to create employment and ensure decent wages and working conditions, in addition to better living conditions based on localizing STI applications. Additionally, there is a need to boost domestic research and development and promote human resources, particularly to reverse migration of skilled labour to other regions, commonly referred to as the brain drain. Based on these new realities, effective regional and international cooperation is the key to tangible and long-lasting results.

Within this context, the Regional Agenda for Action on Technology, Employment and Poverty Alleviation (ATPA) was conceived to facilitate and consolidate partnerships between concerned national, regional and international organizations; to help harness new technologies in order to achieve the objectives of the Millennium Declaration; and to create various technology facilities, including technology parks, incubators and centres of excellence.

Ms. Mervat Tallawy,
Executive Secretary of ESCWA

Mr. Rafik Hariri,
Prime Minister of Lebanon

Mr. Samir Radwan,
Adviser to the Director General, ILO, Geneva

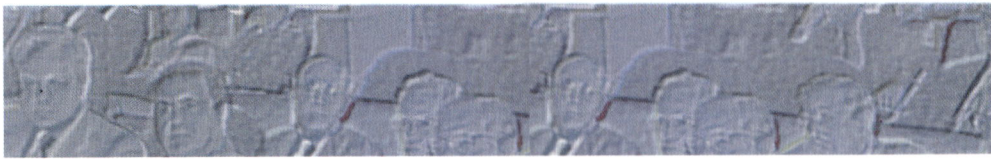

The keynote speech by Professor Zewail

H. E. Prime Minister Rafik Hariri
Ms. Mervat Tallawy Executive Secretary
of ESCWA
Excellencies, Ladies and Gentlemen

I am pleased to address this forum which I believe is timely for identifying what can be done in the Arab world-what can be done for achieving major progress toward the betterment of human life and the active participation in world economy. The language of this century is science and technology (S and T) and as the title of my address indicates, the Arab world has no choice but to develop a new vision for its S and T and to integrate it and implement it as part of the economic, cultural, and even political system.

My personal view is shaped by my origins, experience, and current position. First, as an Egyptian Arab I have genuine interest in seeing progress in this region and I continue to offer my services. Second, I live and work in the United States, and I know how critical the role of S and T is in its progress and that of other developed nations. And, third, my thoughts are not influenced by self-interest - a personal business benefit or political agenda - but rather they are guided by a sense of what is best for the region. With this in mind, here I would like to address the core of the problem and to offer a viable solution, while focusing on the following topics: the Arab world today; the roots of the problem; S and T of the twenty-first century; and concrete steps for a scientific renaissance in the Arab world.

Indeed, the current state of the Arab world calls for a new renaissance to halt the decline in human productivity, world share of new knowledge and freedom, and even regional and international political power. Countries such as Egypt, Lebanon, Jordan, and the Gulf states have built national infrastructures, some at the advanced world level, but along with the rest of the Arab world, they still lack a strong S and T base, without which the Arab world will retain its developing, if not underdeveloped, world status. Because the Arabs possess the human and material resources, we can make the transition to developed world status; moreover, history is on our side. I believe that such a transition is essential, not only for human prosperity and participation in the global economy, but also for peaceful and dignified survival in a world whose currency is knowledge and whose creativity is freedom.

The Arab World Today

Arab history is rich with achievements. Arabs developed a civilization that is polyethnic, multiracial, and international. The Arab and Islamic civilization rose to become the foremost economic power in the world, at which time it also reached the highest level in the sciences-it was a major contributor to the European Renaissance. One must therefore ask, What went wrong in modern times? There are political reasons, both global and regional. Not only did colonization, occupation, and wars impede progress in the Arab states, but also once these

hindrances were out of the way, the new systems did not provide institutions of sufficient democratic nature to utilize the best of human thought and human participation in society. Equally important was the gradual erosion of institutions of knowledge, such as universities, and the human creativity that they nourish.

Numerous analyses of the cause for the decline have been made, and always the critical role of S & T has been emphasized. This year, The Economist reported that the Arabs, who once led the world in science, are dropping even further behind in scientific research and technology. The recent UNDP report concludes that the barrier to progress in the Arab world is not the lack of resources but the lamentable shortage of knowledge, freedom, and the empowerment of women. Going back to the Islamic Ottoman Empire, one historian commented, "it collapsed because it failed to respond to the aggressive European development in science and technology."

Because of this state of the Arab world today, we see alarming trends in productivity, demography, and technical ability exemplified by the following unhealthy symptoms:

▶ (1) Per capita income is among the lowest in the world, at a level just above that of sub-Saharan Africa;

▶ (2) Illiteracy is among the highest in the world, reaching above 50% in some countries;

▶ (3) Youth, who make up more than half of the population of 280 million people in the Arab

world, are unemployed (or misemployed) at an alarmingly high rate-more than 25%.

▶ (4) Per capita participation in global S & T is among the lowest in the world.

Arab Science and Technology

The direct link between a strong S and T base and the progress and prosperity of a nation is not clear to many in the region. In fact, from my own experience, I have found some misconceptions summed up by one common slogan: "S and T is only for the rich countries." In other words, S and T is a luxury. Even worse, some believe that S and T can be bought from developed countries, like other imported products, such as cars and televisions. So how essential is S and T to national progress? A look at the global status of S and T is illuminating. The following statistics highlight the current state of S and T in the Arab world and the role of S and T in the ranking of nations and regions of the world.

The total number of scientific papers published worldwide over the past five years is 3.5 million. According to the Institute of Scientific Information, the percentages of contributions by geographic origin are as follows: European Union (37%), United States (34%), Asia Pacific (21%), India (2.2%), and Israel (1.3%). The contribution from the Arab world, with 280 million people in 22 countries, is less than that of Israel, with a population of some six million people; individual Arab countries' contribution ranges from 0% (Yemen), to 0.03% (typical), to 0.3% (Egypt). Zero percent here means the number of papers is negligible.

These percentages are in the same category with Angola, Nicaragua, Somalia, and Guinea. If these numbers are adjusted to the population of the country, the results are even more striking-the per capita contribution for the

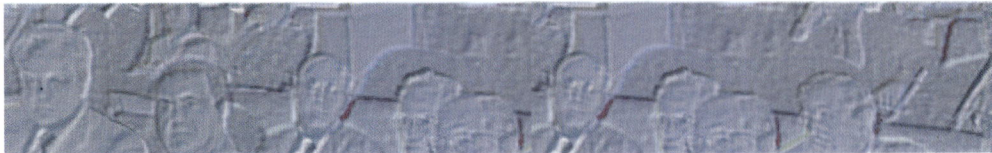

total Arab world is 1% to 2% that of an Israeli, and this refers only to the number of papers without taking into account the impact of research and development. These statistics are not surprising, as there is not one scientific institution in the entire Arab world on a par with the Weizmann Institute or the Technion of Israel, or with similar institutes in India, and certainly not with the Max Planck Institutes of Germany or their like in other developed countries.

One must conclude from these statistics that the current situation reflects the poverty of Arab S & T and the direct connection between the health of S and T and economic progress and prosperity in the world. We note three points:

▶ 1. The strong correlation between the advanced state of S and T and advanced state of the nation. The USA contributes 34% of the world's S and T papers and in parallel contributes 30% - 40% to the world economy.

▶ 2. Developing nations are crossing the line to developed status because of their investment in S and T (including education). Pacific Asian countries are showing exponential growth in S and T papers.

▶ 3. The Arab world is richer than it is developed. It is rich in resources and commodities, but it does not possess the strong base for new knowledge, the S and T base.

Clearly, S and T is not a luxury and cannot simply be purchased; it is essential to progress. Even when it is focused on "luxury" investments-so-called pure research or basic science-it

continues to produce new knowledge, radiate rational thinking, and foster social enlightenment: "al-'ilm nur!" ("knowledge-science-is light!"). Just as important, the power of S and T enhances national pride. I am reminded of the exchange between the physicist Robert R. Wilson, who was instrumental in building the giant atom smasher (Fermi Lab), and Senator John Pastore at a congressional hearing in 1969. Senator Pastore asked: "What is an atom smasher good for? Does it contribute to the security of the country?" Wilson answered: "No, sir. It has no value in that respect. It has nothing to do directly with defending our country-except to help make it worth defending."

Are the Problems Arab?

Some associate the lack of progress in the Arab world with cultural and religious roots. Some even believe that a "clash of civilizations", as termed by Samuel Huntington, is imminent because of the incomplementarity of cultural and religious values between the Arab world and the West. Elsewhere, I have argued that there is no fundamental foundation behind these theories, and that a "dialogue of civilizations" can be realized provided that nations experience economic benefits and observe political fairness; it helps a great deal if ignorance about civilization is fairly unveiled. The examples of dialogue are numerous, and history is the best reference.

For the Arabs, the problems are not in Arab roots or Arab genes. Like all humans, and indeed all living species, they possess the same genetic makeup-their genetic alphabet, G, C, A, and T is the same. Arabs have history to back-up their claim of achievements at the highest level. At present, Arabs working in an appropriate environment excel, even in S and T, now a Western trademark, proving that it is possible for them to peform at the highest level! And Arabs have the human and material resources.

At the moment, what is lacking is a clear system. In the first place the system must serve the collective needs of the population. As every human counts, the foundation of the system for humans, as a species (Homo sapiens, literally, the "rational" or "thinking human"), must be built on two pillars-knowledge and freedom. Education must be reformed at all levels to transform it from an authoritarian teacher-to-student rote transfer to a process that challenges the mind and provides hands-on practical experience. Illiteracy must be eliminated or at least reduced to only a few percent of the population The base for research and development cannot function in the present form and a new vision is needed, as discussed below.

The second part of the foundation involves a fresh system of law, which defines clear civil, cultural, and religious boundaries and applies equally to every person. The main objectives should be the freedom of human thought and the virtual elimination of bureaucracy, the dinosaur crippling the progress

in all sectors. The Arab people are not any less intelligent or capable than the people of the Pacific Asian nations and I believe that the transition to developed status is possible, provided these issues are addressed rationally and collectively.

The Twenty-First Century More Challenges

In the coming fifty years, knowledge-based and skill-based societies will have the lion's share of world market and high status. If there is no new renaissance, extrapolating from the data on the current state of the Arab world, the future seems grim. Without S and T how can the Arabs participate in current world issues such as stem-cell research, cloning, human genome sequencing, artificial intelligence, manipulation of matter, molecular medicine, and cosmology? Without S and T how can the Arabs actively contribute to the world market in technologies such as microelectronics, information and communication, new materials, and the revolutionary biotechnologies?

The challenges require a new system of education and a new outlook on technologies. Technologies fall into three categories, those that are "simple" but relevant to services, solving domestic problems of everyday life, from traffic lights to desalination of water; those that are "innovative," which make participation in the world market possible, such as microelectronics; and those that are "frontier," which are concerned with research into the unknown, representing an investment in the future. To be effective, new education and research and development in the first two categories are required and, at the least, we must be seriously engaged with the issues of the third, frontier category-of where the world is going to

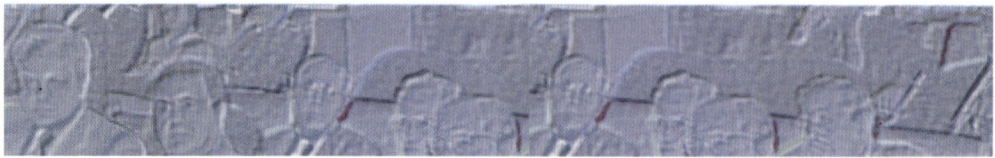

be. I mention here three of the myriad of the new frontiers that countries like Israel and India already have as part of their development:

Our matter-the scale of the very small. We are on our way to being able to manipulate matter at its smallest, most fundamental limits, both in time, on the femtosecond scale, and in length, on the nanoscale. Just think about these new scales of time and space in the world of the very small. If your heart beats once a second, now we can see the beats of atoms in a femtosecond, in a millionth of a billionth of a second-a femtosecond is to a minute like a minute is to the age of the universe. Similarly, we can study matter on the nanometer scale and resolve the atoms in their structures-the size of the atom to the size of the earth is like the size of the earth to the whole universe. The opportunity is huge for acquiring new knowledge and for creating new forms of "our matter." The interface of matter's micro- and nanonetworks, designed to produce artificial intelligence, to our life organs, such as the brain, will be another frontier that could alter the boundaries and meaning of species.

Our universe-the scale of the very big. In this century, we may have colonies on the moon, and we may have our second homes on other planets and maybe even in other galaxies. Just think of the scales of the world of the very big. Our universe is about 12 billion years old, and at the speed of light, which is 300,000 km/s, our universe's limit of distance is 100 billion trillion kilometers-certainly enough space for the six billion people on earth today, even multiplying by ten or by one million for the future! The opportunities involving outer space and information technology are unlimited. Through "virtual walls," which in principle will provide any information one needs, education and intelligence in all societies will be redefined.

Our life-the scale in between. In the first year of this century, the sequencing of the human genome was completed. We now have the genetic map that describes every human on planet Earth. Just think, three billion letters have been deciphered and read into our book of life. The history of biology has changed from the classification of living organisms (Darwin's theory), to the world of cells (Leeuwenhoek's microscope), to now the molecular world (Watson and Crick's DNA). Soon we might see a nanoscale motor entering the cell to do work. Medicine and human health will certainly enter a new age. Likewise, there will be new opportunities in engineering, economics, law, humanities, and other fields. In fact, twenty-first-century knowledge may revert to Aristotle's style of philosophy by emphasizing a "no fragmentation" approach, or what is called nowadays multi-disciplinary science, where many disciplines become interwoven. But with any major development we have to think about society, or more specifically about the transition from scientific discovery in the laboratory to its technological implementation, and to assessment of the benefits or possible harm to society. We have succeeded in the first cloning experiment of a higher organism-Dolly, the sheep-but we still do not have a full perspective of its moral, ethical, and religious implications. These are complex issues for science and society to tackle, and only societies with a scientific and humanistic culture can address these issues rationally.

These new areas of science and technology require a sustained base for research and development. But some clarification is needed. Although some "poor" Arab countries do not possess the means to do frontiers research, even they can undertake reforms of the educational system and select certain technologies that are needed for export of goods to and benefit from world markets. Rich Arab countries are surely capable of supporting frontier S and T, as they do in supporting commodities, and they could set up their institutes to draw talent from the entire Arab world-their

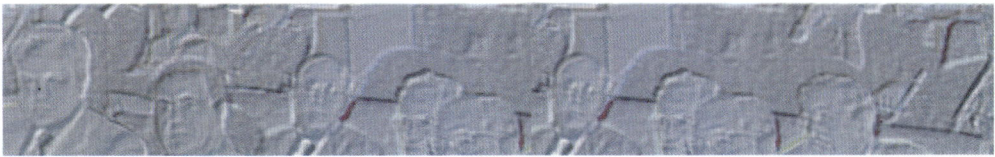

breakthroughs would benefit society and halt the brain drain. There are attempts to build certain technologies such as information technology or biotechnology, but frankly these are empty slogans, manifested primarily in new buildings and new equipment. The S and T base in the Arab world, poor or rich, is weak. For the twenty-first century, the Arab world needs revolutionary changes in education and in building a real S and T base.

Concrete Steps for Renaissance

The following are some of my thoughts and suggestions that form a proposal for making the transition. The elements of the proposal are:

▶ (1) Creating a new system of education. This involves the restructuring of methods of education, emphasizing critical, not authoritarian, rational thinking and building science education with a new perspective on social ethics and the humanities. The objective is to provide a new workforce equipped with twenty-first-century tools of education and skills and with a belief in ethics and teamwork. This will not be accomplished without also changing the status and education of teachers.

▶ (2) Creating new centers of excellence. These centers should be on a par with other world-class centers, but with limited focus on areas especially relevant to the region and to global participation. Elsewhere, I have detailed a plan for these centers in Egypt and the Arab world. These centers should not be established without a clear vision and institutional system, and should not be considered as business entities.

▶ (3) Creating new industries. These industries should be based on "local" developed S and T, not on a mindset for purchase of "foreign" technologies. Technology transfer is encouraged, but without a local base these new technologies will always depend on expertise from the outside.

The industries should have a strong link to a technology park to involve new generations of graduates, as I outlined elsewhere in the structure of the University of S and T. The involvement of the private sector and the elimination of bureaucratic obstacles are key to the success of the new industries.

▶ (4) Creating a national science and technology foundation. There is a need to create a foundation whose purpose is to support S and T research and development solely based on a merit peer-review system. This will help the nation enormously in identifying and supporting the best researchers and will seriously engage different institutions to solve national problems of significance.

▶ (5) Creating the Arab Academy of Sciences. Such an academy will elect the best in the Arab world in S and T and allow for the exchange of knowledge with colleagues around the world. The academy should also have the role of a scientific think tank for the purpose of studying significant national problems and recommending to governments appropriate solutions. The academy must be autonomous.

The above five-point proposal will not succeed without granting independence, with accountability, to S and T institutions, in order to free them from bureaucracy and allow for scholarly debates. Moreover, the proposal will not be effective without shielding the institutions from political influences and fanaticism.

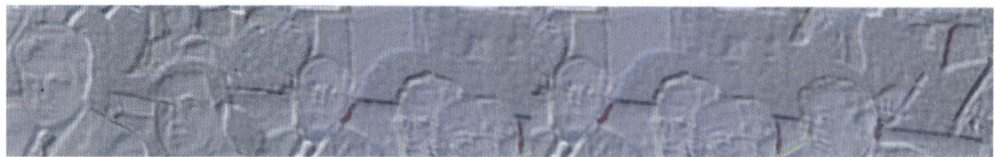

Closing Remarks

Tony Blair in speaking at the Royal Society this year said: "We need to ensure that government, scientists and the public are fully engaged together in establishing the central role of science in building the world we want. We could choose a path of timidity in the face of the unknown. Or we could choose to be a nation at ease with radical knowledge, not fearful of the future, a culture that values a pragmatic, evidence-based approach to new opportunities. The choice is clear. We should make it

Renaissance. The people have the brainpower, and history is on our side. People also have the resources, and in many countries the infrastructure has improved considerably. What we have lacked is a rational system able to transform the promises of a rich culture and to take advantage of the new advances in the world, scientifically, economically, and politically.

True, there are conflicts in the region and there are domestic problems that are draining resources. However, there is no choice for the Arabs if we want to make the transition to advanced-world status but to produce a

confidently." The UK Prime Minister, despite his nation's premier position in the developed world, is concerned about the future of science in building a new world. Arab leaders should be deeply concerned about the state of science and technology in the Arab world, not only its future, but also its present.

By nature, I am an optimist, and because of this optimism I have been involved for more than ten years, in different ways, to help the Arab world. I can see how the Arabs could produce a

renaissance-not an incremental change! S and T is the new currency of the twenty-first century and without achieving new levels of education, skill, and science culture, there will be no possibility for a change. We Arabs must do it ourselves and should not be handicapped by conspiracy theories. We must regain confidence and pride in ourselves. Slogans alone will not work. Only by a new vision and by a new commitment will we Arabs make the transition. History will never forgive this generation if we do not.

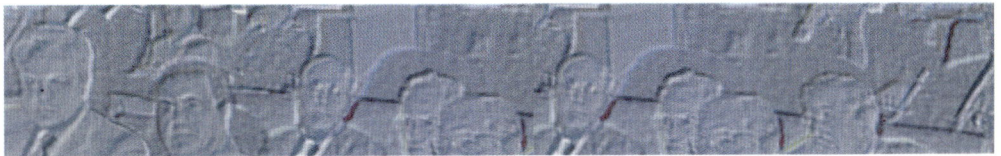

Relevant References

- "Self-doomed to failure", *The Economist* (London, 6 July, 2002), p. 25.
- *Arab Human Development Report 2002*, United Nations Development Program, (New York, July 2002).
- Ahmed Zewail, "Dialogue of civilizations: making history through a new world vision", *Science and the Spiritual Quest/ Science et la quête du sens,* (UNESCO, Paris, 2002).
- Ahmed Zewail, "Science for the have-nots", *Nature* (London, vol. 410, 2001), p. 741.
- Ahmed Zewail, *Voyage through Time - Walks of Life to the Nobel Prize*, (American University in Cairo Press, 2002).

Acknowledgments

The visionary efforts of Ms. Mervat Tallawy, through the leadership of the UN Economic and Social Commission for Western Asia (ESCWA), to bring about this Forum should be appreciated by all of us and I take this opportunity to thank her on behalf of all participants. We also wish to thank Mr. Omar Bizri for his untiring effort in organizing the Forum. I'm indebted to Mr. David Pendlebury for his sincere effort with the S and T data (ISI).

Chapter 2

The United States of America

Opinions-Editorials (Op-Eds)[*]

* *References to these Articles are given in the Index at the end of the book.*

Los Angeles Times

Published: 19 August 2012 (Sunday Edition)

How curiosity Begat Curiosity

Scientific breakthroughs come from investing in science education and basic research.

Op-Ed by Ahmed H. Zewail

On Aug. 5, I was among those who witnessed the rover Curiosity landing on Mars in real time at NASA's Caltech-managed Jet Propulsion Laboratory. The excitement was overwhelming: The one-ton Mars Science Laboratory broke through the Red Planet's atmosphere, slowed its speed from 13,000 mph to almost zero and touched down. One glimpse of those first images from more than 100 million miles away demonstrated America's leadership in innovation.

Kelley Clarke, left, celebrates at JPL as the first pictures appear on screen from Curiosity. (Los Angeles Times / August 5, 2012)

Curiosity — the rover and the concept — is what science is all about: the quest to reveal the unknown. America's past investment in basic science and engineering, and its skill at nurturing the quest, is what led to the Mars triumph, and it is what undergirds U.S. leadership in today's world. But now, decreases in science funding and increases in its bureaucracy threaten that leadership position.

After World War II, scientific research in the U.S. was well supported. In the 1960s, when I came to America, the sky was the limit, and this conducive atmosphere enabled many of us to pursue esoteric research that resulted in breakthroughs and Nobel prizes. American universities were magnets to young scientists and engineers from around the globe. The truth is that no one knew then what the effect of that research would be; no one could have predicted and promised all that resulted. After all, it is unpredictability that is the fabric of discovery.

In much of academia today, however, curiosity-driven research is no longer looked on favorably. Research proposals must specifically address the work's "broad relevance to society" and provide "transformative solutions" even before research begins. Professors are writing more proposals chasing less research money, which reduces the time available for creative thinking. And with universities facing rising costs generally, professors are more and more involved in commercial enterprises, which may not always push basic research forward. Even faculty tenure may be driven less by how good one is at science than how good one is at fundraising.

These constraints and practices raise the question: Would a young Albert Einstein, Richard Feynman or Linus Pauling be attracted to science today? Would they be able to pursue their inquiries into fundamental questions?

A generation ago, at the same time that government was supporting curiosity-based research, so was U.S. industry. One of the jewels was Bell Labs, where fundamental research was so advanced that it used to be said that it was "the best university in America." Bell Labs employed some of the world's leading scientists and engineers, and collectively they made pioneering contributions, from the discovery of the tiny transistor to the "big bang" origin of our universe.

The broad-based fundamental research at Bell Labs is no longer pursued, and other industrial labs have, for the most part, disappeared or redirected their resources into much more product-oriented research.

I teach at Caltech and oversee a research laboratory there. In general, I find that the majority of young people are excited by the prospects of research, but they soon discover that in the current market, many doctorate-level scientists are holding temporary positions or are unemployed. The average age at which principal investigators receive their first major government grant has risen, and experience from multiple postdoctoral positions is often necessary for advancement in academia. This slow track discourages young scientists from pursuing research careers.

So what is the formula for better "managing" discoveries? The answer is in the natural evolution of research and development, from curiosity-driven science to technology transfer and then to societal benefits.

We must nurture creative scientists in an environment that encourages interactions and collaborations across different fields, and support research free from weighty bureaucracies. The nation must provide young people with a proper and attractive education in science, technology, engineering and math. And the best minds from around the world should be encouraged, not discouraged, by public policy to join in this American endeavor. In sum, a renewed vision for investment in fundamental research is needed, especially in Washington, where further cuts across the board in science funding are being contemplated.

In the 1950s, Nobel laureate Robert Solow showed that new technologies create a large portion of economic growth, affecting nearly 75% of the growth output in the U.S. The theory of quantum mechanics alone has had a major impact. Without it, revolutionary technologies would not have been realized. Think of the laser, optical communications, MRI and discoveries in drug design, gene technology and miniaturization. At the same time, American influence in the world is bolstered largely through its "soft" power, and science and technology is an essential force of this influence, according to the Pew Research Center's Global Attitudes Project poll.

Since the Industrial Revolution, the West has dominated world politics and economics with the power of science. Since the mid-20th century, the United States has been at the center of that dominance, and more recently, China is pouring resources into R&D to reach first world status. The U.S. can still maintain research institutions, such as Caltech, that are the envy of the world, yet it would be hubristic and naive to think that this position is sustainable without investing in science education and basic research. We do not know now what will be relevant tomorrow.

American innovation and leadership put the rover Curiosity on Mars. Now is the time to recommit to the wise vision that made it happen — otherwise the sun of innovation will come from the east.

Ahmed Zewail, winner of the 1999 Nobel Prize in chemistry, is a professor of chemistry and physics at Caltech. He also serves on President Obama's Council of Advisors on Science and Technology.

theguardian

Published: 12 July 2010

The US Needs a New Soft Era

By focusing on science and education, America can rebuild its relations with the Arab world.

Op-Ed by Ahmed H. Zewail

Earlier this year I was in Alexandria, speaking about educational reform in front of a packed auditorium of students, teachers, and professionals. I was there as the US president's science envoy to the Middle East. I was surrounded by talented young people, ambitious for themselves and for their country. They represent the hope of Egyptian society and are the ones whom Barack Obama's Cairo initiative, "to seek a new beginning between the United States and Muslims around the world ... based upon mutual interest and mutual respect", must motivate and engage.

I recalled myself at their age, harbouring similar hopes and ideals, and how science shaped my life. My ambition was moulded by the excellent educational system that existed at that time, supported by a society that regarded academic achievement as a national priority. In that climate, science was not perceived as a threat to religion; in fact it was quite the opposite. The mosque was the neighbourhood house of worship, but it was also the place where my highschool friends and I came to study.

Although the Nasser revolution of 1952 was secular, the culture remained deeply religious — but it was a faith of moderation and tolerance. Women made up nearly half my class at university, and my senior academic adviser there was a woman. In Alexandria my friends were Christians and Muslims.

For my generation, America was not exactly seen as our friend. The US was in conflict with Nasser, it denied aid for the construction of the Aswan High Dam, and supplied Israel with its military arsenal. But despite these anti-American feelings, we were drawn to its soft power — the scientific achievements and constitutional values. Even after the six-day war, when relations between the US and Egypt plunged, my university professors, who had earned their PhDs in the US, gave us a more nuanced view of America, and indeed played a critical role in my coming to the US.

In adapting to life in the melting pot of America, I discovered that the same soft power of science has a huge influence in building bridges between cultures and religions — and has the potential to do so with the Muslim world.

By contrast, hard power is very costly. In the latest Iraq war it caused the death and suffering of millions. No matter what "good intentions" the president and the neocons had in mind — be it the spread of democracy or the security of oil supply — the war engendered more conflict in the Middle East, and diverted attention from economic development in the region and a solution to the Palestinian-Israeli conflict.

There is nothing in the cultural DNA of Islam that makes it resistant to assimilating new ideas. The vast majority of Muslims are moderates who want nothing more than to live a decent life and see their children educated. Everywhere I went in Alexandria people expressed eagerness to forge closer scientific and educational ties with the US, whatever their disagreements on political issues.

In this tumultuous part of the world what is needed most is the soft power of modern science, education and economic developments. Close to half of the 300 million Arabs are now under the age of 15, and unemployment is above 15%. This situation is a timebomb that could be triggered by frustrated youth expressing their despair through national and international violence. Progress in the Middle East is important to the West not only for obtaining natural resources, but also for maintaining an influence in a region that is luring other powers such as China and Russia.

For half a century US policy has focused on securing the flow of oil and ensuring Israel's military superiority; it has supported undemocratic regimes while calling publicly for democratic change. This two-faced policy must change to one that genuinely supports human rights and good governance. In the places I visited, people wish to see an even-handedness on Palestinian issues. In the long run the best support the US can give Israel is a secure peace.

We need a long-term and coherent partnership to build up and modernise science, increasing support to students and scholars. The highly qualified Arab diaspora can be involved in this partnership. Surely the aspirations and energies that I encountered in Alexandria and throughout the region can be harnessed, through soft power, to usher in a new era in the relationship between the West and the Arab and Muslim world.

Ahmed Zewail is the Linus Pauling Chair professor at Caltech and the recipient of the 1999 Nobel prize in chemistry. Last year he was appointed to President Obama's council of advisers on science and technology.

Los Angeles Times

Published: 27 June 2010 (Sunday Edition)

Science as a Shaper of Global Diplomacy

The U.S., admired worldwide for its leadership in technology, should pursue science diplomacy with Muslim-majority countries. Such a policy could complement efforts to promote human rights.

Op-Ed by Ahmed H. Zewail

In today's world, America's soft power is commonly thought to reside in the global popularity of Hollywood movies, Coca-Cola, McDonald's and Starbucks. But the facts tell a different story. In a recent poll involving 43 countries, 79% of respondents said that what they most admire about the United States is its leadership in science and technology. The artifacts of the American entertainment industry came in a distant second. In the 1970s, what I, as a young foreign student studying in the United States, found most dynamic, exciting and impressive about this country is what much of the world continues to value most about the U.S. today: its open intellectual culture, its great universities, its capacity for discovery and innovation.

By harnessing the soft power of science in the service of diplomacy, the U.S. can demonstrate its desire to bring the best of its culture and heritage to bear on building better and broader relations with the Muslim world and beyond.

I felt the full force of this soft power when I came to the United States from Egypt in 1969 to begin graduate studies at the University of Pennsylvania. I discovered how science is truly a universal language, one that forges new connections among individuals and opens the mind to ideas that go far beyond the classroom. My education here instilled in me greater appreciation for the value of scholarly discourse and the use of the scientific method in dealing with complex issues. It sowed, then nurtured, new seeds of political and cultural tolerance.

But perhaps most significant was that I came to appreciate the extent to which science embodies the core values of what the American founders called "the rights of man" as set forth in the U.S. Constitution: Freedom of thought and speech, which are essential to creative advancement in the sciences; and the commitment to equality of opportunity, because scientific achievement is blind to ethnicity, race or cultural background.

In January, appointed by President Obama as America's first science envoy to the Middle East, I embarked on a diplomatic tour that took me to Egypt, Turkey and Qatar. I met with officials from all levels of government and the educational system, as well as with economists, industrialists, writers, publishers and media representatives. What I learned during these visits was cause for some alarm, but also for considerable optimism.

The alarming aspect comes from the fact that education in many Muslim-majority countries now seriously lags behind international standards. Deficiencies in education, together with widespread economic hardship and the lack of job opportunities for young people, are sources of frustration and despair in many Muslim societies. They are rooted largely in poor governance and growing corruption, compounded by overpopulation and by movement away from the enlightened education I was fortunate enough to enjoy in Egypt in the 1960s.

Yet there are many positive signs as well. Muslim-majority countries such as Malaysia, Turkey and Qatar are making significant strides in education and in technical and economic development. Egypt, Iraq, Syria, Lebanon, Morocco and Indonesia are examples of countries still rich with youthful talents. Nor is this transfer of wealth and learning flowing exclusively from the West to the East. Today there are many Muslims in the West who have excelled in all fields of endeavor. These accomplishments and the values they represent can help the Muslim world recover its venerable heritage as a leader in science by complementing local efforts and aspirations.

It is certainly in the best interests of the United States to foster relations with moderate majorities who today often find themselves locked in struggle with minorities of fanatics. Most people I met in the Middle East believe in Obama's intentions, as laid out in his Cairo speech last year, and welcome the prospect of enhanced scientific and educational partnerships with the United States. Yet some expressed skepticism, with one high-ranking official asking me, "Will the political climate in the United States, and particularly the U.S. Congress, allow him to follow through on his promises?"

To enhance the prospects for success, we should begin by stressing three points.

First, the United States needs to define a coherent and comprehensive policy for pursuing science diplomacy with Muslim-majority countries. Despite many efforts by both public and private organizations, their initiatives remain fragmented.

Second, the focus of a better-integrated effort should be on improving education and fostering the scientific and technological infrastructure that will bring about genuine economic gains and social and political progress. One way would be for the United States to encourage and support the creation of relatively simple earth science labs in elementary schools, along with the teacher training necessary to stimulate curiosity about the workings of nature. For older students, I propose a new program, "Reformation of Education and Development," whose acronym, READ, would have special significance for Muslims, as it is the first word of the Koran. And through the program, the United States should be a partner in establishing science and technology centers of excellence for talented high school and university students in the region.

Third, these efforts must complement, not replace, U.S. efforts to promote human rights and democratic governance in the Muslim world. The United States must also continue to pursue a just and secure two-state solution to the Palestinian-Israeli conflict and work toward freeing the Middle East from nuclear proliferation.

All these efforts would go far toward creating goodwill, catalyzing progress and redirecting the region's energies into new, constructive and mutually beneficial channels.

The soft power of science has the potential to reshape global diplomacy.

Americans like to say that actions speak louder than words, and action is what we need now.

Ahmed Zewail, the winner of the 1999 Nobel Prize in chemistry and President Obama's science envoy to the Middle East, is a professor of chemistry and physics at the California Institute of Technology.

Published: 17 April 2008

We Need a Science White House

Op-Ed by David Baltimore and Ahmed H. Zewail

Tomorrow Hillary Clinton, Barack Obama and John McCain should have been going toe-to-toe in a televised science debate. All three were invited by a bipartisan group of Nobel laureates and other scholars called Science Debate 2008 to step on stage at the Franklin Institute in Philadelphia and explain how they will ensure that America continues to dominate the sciences. Leading in scientific research and advancement is an essential element to our future prosperity, health and national defense.

All three candidates declined. Apparently the top contenders for our nation's highest elective office have better things to do than explain to the public their views on securing America's future.

Protecting that future starts with understanding that much of the wealth in this country comes from scientific research and technological innovation. Translating science into commerce has opened up vast new fields of endeavor and has raised the standard of living in America. The country that is on the cutting edge of developing new technology is the country best positioned to benefit from that new technology.

A clear example is biotechnology. The U.S. is a leader here, and is able to capitalize on its preeminence with disease-resistant crops, anticancer drugs and much more. By developing a strong understanding of the basic science that underlies advances in biotechnology, we are also creating a good training ground for a future generation of scientists and innovators.

But America cannot simply assume its lead in science will continue. In recent years the science community has been starved of the resources it needs. Young, new, energetic scientists are the seed corn of nearly all new scientific development. However, our schools, laboratories and granting agencies all, in one way or another, discourage launching a career in the sciences. There are few grants to live on; and both schools and laboratories have long since lost the sense of joy we remember from our younger days. Science can be exciting and attractive. But convincing bright students to become scientists requires a lot more than we are now providing.

A young university scientist today spends much of his or her time scouring up funding rather than wrestling out the secrets of nature. And the young are not so young. At the

National Institutes of Health, the average age of a first grant is 42 for a Ph.D. and 44 for an M.D. We need policies that nurture excellence and give scientists independence at a younger age. And we need to make American science attractive to both those who were born here and those who were born abroad.

Last year things seemed hopeful, at least for the physical sciences. The National Academy of Sciences issued a report, "Rising Above the Gathering Storm," that helped drive Congress to pass legislation — the American Competitiveness Initiative (ACI) — aimed at bolstering the sciences. It was supposed to beef up the study of science in high school. In the end, no money was found to fund the initiative. It was a commitment made, but not kept.

That's embarrassing as well as shortsighted. We need to re-energize our commitment to being the world's leader in science and technology. We can start doing that by doing a few things:

We need a president who moves science back into the White House. Today we do not have a presidential science adviser and there is no office of science in the White House.

Our government needs to treat science honestly. When the world's scientists flag global warming as a threat to our way of life, it is a warning that should be taken seriously. Stewardship of the planet is our responsibility. No one else is going to do it for us.

We need to fund ACI and double the National Science Foundation's budget for basic research. The government should fund science at a level that will ensure that the U.S. stays in a leadership position in areas like biotechnology, military preparedness, electronics and communication. We need to pay special attention to health research.

We also need to encourage young people to become educated about scientific issues, regardless of whether they become scientists.

This would all be a start. But a complete overhaul of national science policy is needed to prepare the U.S. for a future rapidly overtaking us. Our presidential hopefuls should be telling us their positions on critical science issues, but they have not done so yet. We hope they become more responsive in the months ahead.

Mr. Baltimore, a professor of biology and president emeritus at the California Institute of Technology, was awarded the Nobel Prize in biology in 1975. Mr. Zewail, a professor of chemistry and physics at the California Institute of Technology, was awarded the Nobel Prize in chemistry in 1999.

Essays and Treatises*

WORLD VIEW *A personal take on events*

Curiouser and curiouser: managing discovery making

Beware the urge to direct research too closely, says Nobel laureate **Ahmed Zewail**. *History teaches us the value of free scientific inquisitiveness.*

On a recent official visit to southeast Asia, a prime minister asked me: "What does it take to get a Nobel prize?" I answered immediately: "Invest in basic research and recruit the best minds." This curiosity-driven approach seems increasingly old-fashioned and underappreciated in our modern age of science. Some believe that more can be achieved through tightly managed research — as if we can predict the future. I believe this is an unfortunate misconception that affects and infects research funding. I hear repeatedly, particularly in developing countries: "Applied research is what we need." There is nothing wrong with a nation having mission-oriented research and development to solve specific problems or to dedicate to an outreach programme, such as space exploration or alternative energy. During my visits as a US science envoy, I have emphasized that without solid investment in science education and a fundamental science base, nations will not acquire the ground-breaking knowledge required to make discoveries and innovations that will shape their future.

There are countless examples of breakthroughs from research driven by curiosity. In my first year as a faculty member at the California Institute of Technology (Caltech) in Pasadena in 1976, the late Richard Feynman and I spoke about a theoretical paper he published 20 years earlier in quantum optics that opened the field to experiments, offering a way to visualize the interaction of laser light and matter. With a smile he told me that at the time he only wanted to answer a fundamental question: if a spin moment can precess in a magnetic field, would an optical transition moment do the same? Perhaps a more common example is the development of the laser by Charlie Townes. At the 50th-anniversary celebrations of its invention in Paris this summer, Townes noted that he was driven at the start only by fundamental questions on microwave spectroscopy and how to amplify light. As I told the audience in Paris, it was curiosity that brought about my contributions to femtosecond science, for which the Nobel prize was awarded, and to four-dimensional electron microscopy to visualize matter in space and time.

Quantum mechanics, relativity, and the deciphering of the genetic code are discoveries made along the same paths, as are revolutionary technologies such as magnetic resonance imaging (developed from curiosity-driven research about the spin of an electron) and the transistor (discovered as a result of curiosity about the nature of electrons in semiconductors). The manufacturing, medical and digital-information technology industries that followed now constitute the backbone of global communications and the economy. Curiosity pays!

How can we ensure that such research is encouraged today? Curiosity-driven research requires that creative scientists work in an environment that encourages interactions between researchers and collaborations across different fields. But such attributes cannot and should not be orchestrated by structured and weighty management, as creative minds and bureaucracies do not work harmoniously together. So is there a formula for managing discovery making? The answer lies in accepting a triad of essentials. First, and most important, are the people involved. Giving proper priority to providing thorough and inspiring education in science, technology, mathematics and engineering is essential. Research and development needs to attract the best young minds. Large buildings and massive funds will not produce much without the right people.

Second, an atmosphere of intellectual exchange is of paramount importance for ideas to crystallize. To distract faculty members with the writing of extensive and numerous proposals or to turn them into managers is the beginning of the end. The modern enterprise of science has become so bloated and complex that the traditional models of funding must be re-examined. How do we focus resources on the best science and what is the level of funding needed to serve society best?

Third, without resources little can be achieved, no matter how creative the mind. Obviously, investment in science is needed to build instruments and to hire competent staff. Countries and institutions that provide the requisite infrastructure and the funding for ideas will be the homes of discoveries. But such support should follow the vision of creative researchers, not be built merely to lure money or to force people into fashionable research areas such as nanotechnology.

Today, officials in many developing countries want to find ways to reach the innovation levels of the developed world. In the search, they often overlook the key roles of fundamental research and science education; regrettably, the same trend is creeping into developed countries. Political leaders must appreciate that it is the quest for new knowledge that drives innovation, and without it young students will not be attracted to the profession.

I have been fortunate to spend the past 30 years of my career in an institution that believes in such values. Despite pressure to change, I hope that Caltech will continue to preserve its unique culture as — in the words of one of my colleagues and former Caltech president, David Baltimore — a "village of science". Preserving knowledge is easy. Transferring knowledge is also easy. But making new knowledge is neither easy nor profitable in the short term. Fundamental research proves profitable in the long run, and, as importantly, it is a force that enriches the culture of any society with reason and basic truth. ■

MAKING NEW KNOWLEDGE IS NEITHER **EASY** NOR **PROFITABLE** IN THE SHORT TERM.

↻ NATURE.COM
Discuss this article
online at:
go.nature.com/lyddyp

Ahmed Zewail *won the 1999 Nobel Prize in Chemistry. He serves on Barack Obama's Council of Advisors on Science and Technology.*
e-mail: zewail@caltech.edu

The Soft Power of Science

Ahmed Zewail

S cience, the driving force behind so much in the modern world, has rarely figured as a tool for forging and advancing relations among nations. But this has begun to change. Last June, President Obama laid out a blueprint for a "new beginning" with the Muslim-majority countries in a major speech at Cairo University. The President expressed optimism about creating strong, enduring ties rooted in common interests and mutual respect. In particular, he called for scientific and educational collaborations that could both cement those ties and serve as engines of social, economic and political progress.

The President's initiative must be followed with an action plan, and the United States is in a good position to do so. Much of America's global influence is based on its leadership in science and technology, and so the United States would do well to integrate the "soft" power of American science into its diplomacy.

My own experience as a product of both the East and the West has taught me how formidable the "soft" power and transformative cultural potential of science can be. As a young man growing up in Egypt, I did my undergraduate work in Alexandria, a city steeped in the history and culture of science and a cosmopolitan center in a Muslim-majority nation. Its population

was ethnically and religiously diverse, with Muslims and Christian Copts, as well as Arabs, Greeks, Italians and others living peacefully side by side. Women made up nearly half of my class at the University of Alexandria; indeed, my senior research adviser was a woman. Religious Egyptians of all denominations enthusiastically embraced arts, literature, theater and music. I cannot recall a single incident of terrorism by religious fanatics. It was this soft power, these values and this culture, embedded in and supported by strong educational and media systems that, far more than weapons and political hegemony, constituted Egypt's chief export to the rest of the Arab world and the source of its leadership throughout the region.

In today's world, America's soft power is commonly thought to reside in the global popularity of Hollywood movies, Coca-Cola, McDonald's and Starbucks. But the facts tell a different story. In a recent poll involving 43 countries, 79 percent of those surveyed said that what they most admire about the United States is its leadership in science and technology. The artifacts of the American entertainment industry came in a distant second. What I, as a young foreign student in the 1970s, found most dynamic, exciting and impressive about the United States is what much of the world continues to value most about America today: its open intellectual culture, its great universities, its capacity for discovery and innovation.

I felt the full force of this soft power when I came to the United States in 1969 to begin graduate studies at the University of Pennsylvania. I discovered how science is truly a universal

Ahmed Zewail, *the Linus Pauling Chair Professor at the California Institute of Technology, received the 1999 Nobel Prize in Chemistry. He serves on President Obama's Council of Advisors on Science and Technology, and as U.S. Science Envoy to the Middle East.*

language, one that forges new connections among individuals and opens the mind to ideas that go far beyond the classroom. My education in America instilled in me greater appreciation for the value of scholarly discourse and the use of the scientific method in dealing with complex issues. It sowed, then nurtured, new seeds of political and cultural tolerance.

But perhaps most significant was that I came to appreciate the extent to which science embodies the core values of what the American Founders called "the rights of man" as set forth in the U.S. Constitution: freedom of thought and speech, which are essential to creative advancement in the sciences; and the commitment to equality of opportunity, because scientific achievement is blind to ethnicity, race or cultural background.

By harnessing the soft power of science in the service of diplomacy, America can demonstrate its desire to bring the best of its culture and heritage to bear on building better and broader relations with the Muslim world and beyond. In January, as America's first Science Envoy to the Middle East, I embarked on a diplomatic tour that took me to Egypt, Turkey and Qatar. I met with officials from all levels of government and the educational system in these countries, as well as with economists, industrialists, writers, publishers and media representatives. What I learned during these visits was cause for some alarm, but also for considerable optimism.

The alarming aspect comes from the fact that education in many Muslim-majority countries now seriously lags behind international standards. Deficiencies in education, together with widespread economic hardship and the lack of job opportunities for young people, are sources of frustration and despair in many Muslim societies. They are rooted largely in poor governance and growing corruption, compounded by overpopulation and by movement away from the enlightened education I was fortunate enough to enjoy in Egypt in the 1960s.

Yet there are many positive signs as well. Muslim-majority countries such as Malaysia, Turkey and Qatar are making significant strides in education and in technical and economic development. Egypt, Iraq, Syria, Lebanon,

Morocco and Indonesia are examples of countries still rich with youthful talents. Nor is this transfer of wealth and learning flowing exclusively from the West to the East. Today there are many Muslims in the West who have excelled in all fields of endeavor, from science, technology and business to arts and the media. These accomplishments and the values they represent can help the Muslim world recover its venerable heritage as a leader in science by complementing local efforts and aspirations. Ultimately, of course, Muslims themselves must be responsible for their own destinies, but Muslim countries possess a wealth of human and natural resources that must not be wasted. Partnerships between America and other Western countries with Muslim-majority societies can help to tap these reservoirs to build modern, prosperous and creative nations.

It is certainly in the best interests of the United States to foster relations with moderate majorities who today often find themselves locked in struggle with minorities of fanatics. Looking ahead, it is therefore of some importance that the majority of people whom I met during my weeks in the Middle East believes in the sincerity of President Obama's intentions and welcomes the prospect of enhanced scientific and educational partnerships with the United States. Yet some expressed skepticism. "Mr. Obama made a fine speech in Cairo", one high-ranking official said to me. "But will the political climate in the United States, and particularly the U.S. Congress, allow him to follow through on his promises?"

I believe it is important to overcome this skepticism with action. America has the power to bring about genuine change through science diplomacy, to utilize the enormous soft power reserves of its scientific community to establish lasting influence and mutual benefits in an interconnected world. We should begin by stressing three points that will enhance the prospects for success.

First, the United States needs to define a coherent and comprehensive policy for pursuing science diplomacy with Muslim-majority countries. As things stand today, despite there being a science and technology adviser to the Secretary of State, U.S. efforts in this area remain fragmented, underfunded and thus less than the sum of their parts. The assets of the U.S. government in this regard are considerable, residing, for example, within the National Academy of Sciences, the National Science Foundation, the National Institutes of Health, the Centers for Disease Control and Prevention, the Food and Drug Administration, as well as in the Departments of Education, Interior, Agriculture and State. We need to bring these assets into concert. In this regard, too, foreign aid distribution needs to be revisited.

Second, the focus of a better-integrated effort should be on improving education and fostering the scientific and technological infrastructure that will bring about genuine economic gains and social and political progress. One way to build human capital in science, for example, would be for the United States to encourage and support the creation of relatively simple earth science labs in elementary schools, along with the teacher training necessary to stimulate curiosity about workings of nature. For older students, however, I propose a new program, "Reformation of Education and Development", whose acronym, READ, would have special significance for Muslims, as it is the first word of the Quran.

Through the READ program, the United States would support the establishment of centers of excellence in science and technology that can serve as educational hubs for talented high school and university students throughout the region. During my recent trip, I encountered widespread enthusiasm for this idea as well as pledges of support from wealthy Arab nations. Such centers would play a critical role in instilling regional pride and creating the institutional basis for forward-looking, knowledge-based economies. They would serve as tangible evidence of America's commitment to partnership and would help promote peace and stability throughout this part of the world.

Third, these efforts must complement, not replace, U.S. efforts to promote human rights and democratic governance in the Muslim world. The United States must also continue to pursue a just and secure two-state solution to the Palestinian-Israeli conflict and work toward freeing the Middle East from nuclear proliferation. These efforts, which should run parallel with READ, would go far toward creating goodwill, catalyzing progress and redirecting the region's energies into new, constructive and mutually beneficial channels. Focusing on the benefits to majorities will reduce the risks from dangerous minorities and the ravages of terrorism.

The soft power of science has the potential to reshape global diplomacy. If the vision that President Obama set forth in Cairo can be realized, history may one day record that speech as ushering in a period of transformative change. Americans like to say that actions speak louder than words, and action is what we need now. 🐾

A better knowledge and appreciation of the values embedded in scientific inquiry are essential for a liberal society. . . . The basic value system of organized science—its devices for encouraging systematic public criticism, its commitment to seeking out problems and solving them, its communitarian tradition of sharing its findings freely and openly— . . . can serve as a resource to reinvigorate our dedication to public reason. . . . It was the philosopher Karl Popper who best described the commonalities between scientific inquiry and what he called the "open society." Both require traditions that foster an arena where people can voice disagreements and scrutinize conflicting views. If dissent is the soul of democracy, so is criticism the motor of scientific research.

—**Noretta Koertge,** *Scientific Values and Civic Virtues* (2005), p. 4

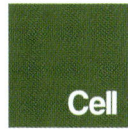

Leading Edge

Commentary

Science in Diplomacy

Ahmed H. Zewail[1,2,*]
[1]Linus Pauling Chair Professor of Chemistry and Physics, California Institute of Technology, Pasadena, CA 91125, USA
[2]President's Science Envoy to the Middle East
*Correspondence: zewail@caltech.edu
DOI 10.1016/j.cell.2010.04.002

Throughout human history, science and technology have been the backbone of innovations that have driven economic development. Yet, rather oddly, they have not been seriously invoked in the pursuit of diplomacy. This Commentary examines the important role of science in diplomacy and its soft-power in world affairs and peace.

The world goes through economic, political, and religious turbulence, but throughout history science has maintained a steady impact on improving human lives. Despite its central role in driving world markets and as a universal language of communication, science has not been seriously invoked in diplomacy. The Science Envoy program outlined by President Barack Obama in his historic Cairo speech last year has the potential to redefine the role of science in the landscape of diplomacy in general and to define a new beginning with the Muslim world of 1.5 billion people. "Scientopolitics" is a term that may become part of the lexicon of diplomacy. Here, as President Obama's Science Envoy to the Middle East, I reflect on the issues involved and the lessons learned so far.

World Order

The world before World War I, described as "La Belle Époque," was one of peace and prosperity. The material standard of living was on the rise, democracy flourished, and continents were connected by railroads, steam ships, automobiles, airplanes, the telegraph, and the telephone. Man conquered the last uncharted territories of the world, and the United States continued to be the land of promise for millions. Achievements in the sciences, literature, and peace were honored with the first Nobel Prizes in 1901. Science and technology played a pivotal role in this progress.

In the 1990s, the world looked beautiful again: European economic and political cooperation took on a new dimension with the creation of the European Union (EU), and China, Japan, and other so-called Asian Tigers assumed a major role in world economic development. In 1989, the unification of Germany gave the world new hope for solidarity with the end of an era of separation. The policy of apartheid in South Africa was abandoned and the world of the Cold War and nuclear armament appeared to have changed into a "world of globalization." Once again, science and technology were a driving force. Information technology brought the world much closer in distance and in time, prompting Thomas Friedman to title his book, "The World is Flat." Science discoveries and innovations transformed the human condition with improvements in communication and health and led to revolutionary developments, including direct visualization of the nano-world, telescopic observations of the very distant, and robotic landings on Mars. That is not to say that the world has become perfect—conflicts still rage, diseases take their toll, and human rights violations continue. Yet the nations of the world are aiming for unity through understanding and cooperation.

World of the Have-Nots

Despite this progress, the distribution of wealth is skewed. Only 20% of the world's population enjoys the benefits of life in the developed world, and the gap between the haves and have-nots continues to increase. The World Bank has reported that out of the 6 billion people on Earth, 4.8 billion live in developing countries, 3 billion live on less than $2 a day, and 1.2 billion live on less than $1 a day, the absolute poverty standard. About 1.5 billion people have no access to clean water and suffer the consequences of water-borne diseases, and about 2 billion people are still awaiting

the benefits of the industrial revolution. The per capita gross domestic product (GDP) in some developed Western countries is ~$50,000, compared with ~$1,000 or less per year in underdeveloped countries.

This vast difference in living standards between the haves and have-nots ultimately contributes to dissatisfaction, violence, and racial and ethnic conflicts. Politically, there are other factors that kindle the frustration of the population of the have-nots. Among them are the double standards that often arise in the resolution of international disputes and in the support of undemocratic or even corrupt regimes for the sake of national economic or political gains. It is not fair to simply blame the haves for the problems of the have-nots, as there are intrinsic political, economic, and cultural issues involved. However, it is in our own interest to optimize world peace through mutual benefit. Can science play a role?

World Partnership through Science

It is clear that world order requires a coherent and comprehensive policy of partnership, especially between the developed and developing worlds. In my view, education and science should be a cornerstone of this policy as they are of paramount importance in achieving progress and prosperity. They are also the tools that can facilitate the alliance between cultures and nations because science is an international language that is not colored by race or culture. Educating the world's children is not impossible and doing so would open new doors for economic opportunities, involvement in democratic governance, and building knowledge-based societies.

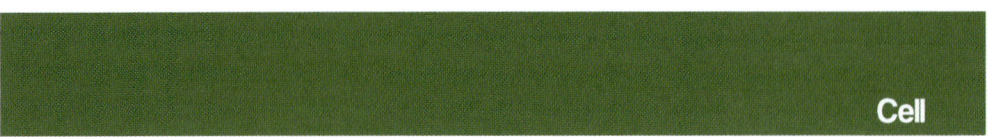

The developed world is so because of its scientific and technological power. The contribution of the U.S. to the world's annual economic output is about 30%, comparable to its share of scientific output on a global scale. Europe's annual economic output is similar and also correlates with its scientific and technological contributions. It is unlikely that this correlation is coincidental. In this century, knowledge-based societies will capture the lion's share of the world economy. As the former UN Secretary General Mr. Kofi Annan pointed out, "Ninety-five percent of the new science in the world is created in countries comprising only one-fifth of the world's population." And, he adds, "Much of that science—in the realm of health, for example—neglects the problems that afflict most of the world's people."

If we are aware of these trends and understand the problems that stand in the way of progress, why do such difficulties and chasms exist in building scientific capacity in the developing world and in harnessing science to improve economic well-being? Admittedly, developing countries have their own political and cultural challenges and responsibilities, but the answer to the question, in part, comes from the fact that science has not been center stage in the foreign policy of developed countries. The United States, through its Agency for International Development (USAID) and other agencies, does provide critical funding of various projects in developing nations, but a new way of thinking about the involvement of science in diplomacy is needed.

Obama's Science Envoys

In his Cairo speech on June 4th last year, President Obama spoke of a "new beginning" with Muslim-majority countries. Besides the relevant political issues, the President addressed issues pertinent to development, education, and science and technology. Specifically, the initiative includes (1)

Figure 1. A New Beginning
In his speech last year in Cairo, President Obama promised to use science and technology in diplomacy as part of a new beginning with the Muslim world. Several major initiatives, including the launch of a Science Envoy program, are already underway. Image courtesy of Al Shorouk Newspaper; art by Hussein Gobayel.

the launching of a new fund to support technical developments and help transfer jobs to the marketplace; (2) the opening of "centers of excellence" for scientific and technological developments in Africa, the Middle East, and Southeast Asia; and (3) the appointment of Science Envoys to build new partnerships and to identify new opportunities for cooperation between the U.S. and Muslim countries.

U.S. Secretary of State, Mrs. Hillary Clinton, announced the Science Envoy program on November 3, 2009 in Morocco. I was asked to be the U.S. Science Envoy for the Middle East; my colleagues Elias Zerhouni and Bruce Alberts are the Envoys to North Africa's Maghreb region and some Gulf countries and to countries in Southeast Asia, respectively. In January this year, I visited Egypt (the most populous country in the Arab world with 80 million people, and a GDP of $2,500 per capita), Turkey (75 million people of Middle Eastern but non-Arab descent, and a GDP of $10,000 per capita), and the Gulf State of Qatar (1.5 million people with 0.3 million Qataris, and a GDP of nearly $100,000 per capita). In these countries, the meetings were wide-ranging and included visits with government officials (heads of state, prime ministers, ministers, and some members of parliament),

members of the education sector (teachers, students, and university professors), institutions of higher learning and research (private and state universities), members of the private sector (economists, industrialists, writers, and publishers), and some media representatives.

The results from these visits are alarming but, at the same time, suggest a new opportunity for the U.S. to address key issues in foreign aid and in partnerships through "science in diplomacy." The alarming aspect comes from the fact that education in many Muslim-majority countries is lagging behind international standards, and universities, including those that led the region (such as Cairo University), are no longer among the world's top 500 according to the Shanghai university ranking system. The current situation is in sharp contrast to the system of schools and universities that existed in the 1960s, when I benefited personally from an excellent education in Egypt. The low performance in world-ranked education, and in the world market, together with the economic hardship felt by the majority and the lack of job opportunities for the youth are becoming sources of frustration and despair. This state of affairs is the consequence of political and social problems, compounded by the increase in population and by the ascending rigidity toward liberal education.

On the positive side, some Muslim-majority countries, such as Turkey and Malaysia, are making significant progress in education. Egypt is rich in human resources and still is home to the best R&D in the Arab world. In both Turkey and Malaysia, I saw the fruits of an improved education system and visited some rising R&D institutions. Even though this progress may be associated with select institutions and does not necessarily reflect the totality of the school and university system in the country, nevertheless the impact on their GDP is evident. Such trends affirm that, in today's world, Muslim-majority

Cell

countries are capable of achieving progress, in this case with supporting efforts from the EU and Japan.

At present, the image of the U.S. in Muslim-majority countries is enjoying a positive trend. Indeed, the speech by President Obama in Cairo has transformed the perception of Muslims at large, and the majority now believes in his good intentions for a new beginning with the U.S. (Figure 1). Some of the people I met are aware of the complexity of America's political system, and they think the actual deliverables, which involve decisions by the U.S. Congress, will not in the end reflect the President's vision. The announcement of the Science Envoy program was well received as a concept in partnership, but again some were skeptical and raised the question: Will this program bring about a paradigm shift in partnership and policy? In spite of these reservations, I believe we have real opportunities to invoke science in diplomacy, to promote new standards of education, and to aid development in the Muslim world.

Science and Technology in Diplomacy

It is remarkable that during all of the meetings conducted for the Envoy mission, there was unanimous agreement on the need for S&T development and in the hope for a leadership role by the U.S. The U.S. is still admired by the youth and by institutions of learned societies and the private sector. Despite various opinions I heard about U.S. foreign policy in regard to political issues of concern in the Middle East, there was full agreement on the importance of science in diplomacy. In this regard, I believe the U.S. should utilize one of its best currencies—science—for foreign policy. Current efforts by many government agencies (the National Institutes of Health, the National Science Foundation, and others) and by private foundations are fragmented. The USAID program provides support in numerous areas and the work is undoubtedly leading to improvements, but the landscape of foreign aid should be redrawn with greater focus on education and on free market economies and job opportunities. The U.S. should also strengthen the role of embassies by appointing

scientists of high caliber as Science Attaches, who can truly mediate in such projects.

Partnerships

Given the state of education and science in Muslim-majority countries, partnerships with the U.S. must transcend the issue of "giving money." Equally important to the funding of projects is the involvement of the U.S. in management and in building the capacity of human resources and infrastructure. This requires a new type of partnership that goes beyond workshops and sporadic exchanges by officials. Perhaps a "Sabbatical System" for U.S. scientists and other professionals may be implemented in order to facilitate in-field management aid and the execution of prototypical projects.

Other rising world powers are involved in creating partnerships and aid programs in the region; for example, the EU is partnering with Turkey, and China is building partnerships in Africa and the Middle East. It is in the best interests of the U.S. to maintain a strong influence in the Middle East, and education and science are tools that can be used effectively to forge diplomatic ties and to build strategic alliances in long-term, profitable partnerships. Through diplomacy, it is also possible to obtain significant funding for such partnerships from economically well-to-do countries in the Muslim world.

Centers of Excellence

As President Obama noted in his Cairo speech, centers of excellence are significant not only for the education of youth but also for the vital growth of any economy. Two prominent examples—India's Institutes of Technology (IITs) and South Korea's KAIST (Korea Advanced Institute of Science and Technology)—have had such an impact and both benefited from U.S. involvement. In the Middle East, we already have a few examples that demonstrate an impact on undergraduate education. The American University in Cairo, on whose Board of Trustees I serve, and the American University of Beirut, which I have visited, are two outstanding institutions that have become a critical source for capacity building, with their graduates becoming leaders in different professions. These institu-

tions, which are standing testimonies of U.S. partnership efforts, also play an important role in cultural exchanges and knowledge enrichment.

Another impressive institution I visited was the Naval Medical Research Unit No. 3 (NAMRU-3) in Cairo, which contributes to U.S. government relations in the region through cooperative research in Egypt and neighboring countries. The NAMRU facilities conduct infectious disease research, including work on the H1N1 influenza virus, with mostly Egyptian personnel in partnership with American professionals residing in Cairo. The infrastructure, including the medical library, has provided substantial aid to other institutions in the country and the region since World War II.

In my view, the U.S. should have a sustainable, long-term plan for creating such centers of excellence, particularly those that can build the foundations for an efficient transfer of knowledge to the marketplace. In each country I visited I received proposals for such centers, and, with focused and directed effort, the U.S. can have a transformative impact in this part of the world. As importantly, partnerships with the U.S. will enable these centers, through legally binding government-to-government agreements, to be essentially free of bureaucratic constraints and regulations, thus providing the free intellectual atmosphere needed for innovation.

Epilogue

When I was a boy growing up on the banks of the Nile, my contemporaries and I viewed America's success story, including its impressive and seemingly unstoppable advances in science and technology, as almost exclusively rooted in the U.S. economic and political system. It was only after moving to the U.S. in 1969 that I came to appreciate the extent to which progress in any society is deep-rooted in that society's cultural values and in its ability to accommodate both reason and faith.

Islam in its pristine state is not a source of backwardness. It was in the Muslim world centuries ago that great civilizations emerged, that world-class universities and scholars arose, and that the rich heritage of ancient Greece and Rome was honored, preserved, and ulti-

mately transmitted to future generations, paving the way for the Renaissance in Western Europe. Today, there are many Muslims in the West who have excelled in nearly all fields of endeavor. If my experiences as a product of both "East" and "West" have taught me anything, it is that the quest for new knowledge and the acquirement of good governance are what lead to progress. It is these values that the Muslim world must cultivate if it is to recover its heritage and take its place among the modern family of nations. Although many Muslim countries possess a wealth of both human and natural resources, it is clear that a cultural rebirth is badly needed—Muslims are ultimately responsible for their own destiny.

Mr. Obama's presidency has the potential to signal a new beginning. I believe that significant progress can be achieved and wish to suggest three pillars for a new policy vision.

First, a coherent and comprehensive, not fragmented, policy for science diplomacy in Muslim-majority countries must be established. The involvement of "science in diplomacy" should focus on education and S&T that are relevant to capacity building and economic progress. I suggest a new program, "Reformation of Education and Development," whose acronym, READ, would have special significance for every Muslim, as it is the first word of the Quran.

Second, a partnership is needed to enable the creation of prototype centers of excellence (in the Middle East and other regions) that will serve as a network among countries. I have discussed possibilities with—and obtained commitments from—several heads of state. Such centers are critical for restoring pride and building new knowledge-based economies and can be a source of enlightenment. They will also stand as testimonies of U.S. partnership and will aid in the peace process.

Third, an unwavering commitment to the issues of human rights and good governance—America's constitutional values—must be made. Today in many Islamic nations, people are demanding change, but their hands are often tied by restrictive, even punitive internal policies. We also badly need concrete steps toward a resolution of the Palestinian-Israeli conflict. This political action, which can be charted in parallel with READ, will catalyze the desired progress by inspiring people to rechannel their energies into creating forward-looking and economically productive societies.

President Obama's election was celebrated in many parts of the Muslim world, and people are hopeful for a new beginning. If the President's vision of cooperation, hope, and mutual respect as set forth in the Cairo speech is followed up with tangible deliverables, it could enter the lexicon of history as the Zero Hour, or as we say in Arabic, "Saat El Sifr," for ushering in an era of real positive change in the world of 1.5 billion Muslims and beyond.

ACKNOWLEDGMENTS

I wish to acknowledge the enthusiastic support of my colleagues on the President's Council of Advisors on S&T, particularly John Holdren, Harold Varmus, and Eric Lander. I thank Jason Rao and Steve Fetter of the White House (OSTP), Bill Lawrence and Manu Bhalla of the State Department, and the Ambassadors and Embassy staff for all their efforts, especially during my 5 weeks of travel.

Chapter 3

The World

Global Peace, Global Education[*]

Angewandte Chemie International Edition

Future of Science and Society

Science for the "Haves"

Ahmed Zewail* and Maha Zewail

1999

More than a decade ago, one of us wrote a commentary titled *"Science for the Have-Nots"*.[1] The author (A.Z.) was and continues to be concerned about the unhealthy state of education and science in the developing world. It was advocated that major reforms of the governing systems, aided by a new type of partnership between the developing world and the developed one, is needed in order to change the plight of the 80% world population of have-nots. In this Essay, on this special occasion, we raise a concern about the state of education and science in the countries of the haves—the developed world—and we believe that reform in this case simply requires a return to the policy of wise investment in basic science, the endless frontier.

Only one-fifth of the world's population enjoys the benefit of life in the developed world and the gap between the haves and have-nots continues to increase, threatening stability. The World Bank informs us that of the 7 billion people on Earth, 5.75 billion live in developing countries; 2.4 billion live on less than US$2 a day; 0.81 billion people still do not have access to an improved water source; and 1.28 billion live on less than $1.25 a day, which defines the absolute poverty standard. Additionally, 1 billion people suffer from hunger today.

Globalization, in principle, is believed to help nations prosper and advance, but in reality, it is better tailored to the fraction of the world's population able to exploit natural resources and markets. The disparity is huge. The per capita gross domestic product—the total unduplicated output of economic goods and services produced within a country as measured in monetary terms—has reached $50000 in some Western countries compared to about $1000 in many developing countries, and significantly less in underdeveloped populations. According to World Bank statistics, Egypt's GDP is US$2800 and South Africa's is US$8100 while South Korea is at US$22400, and the United States is at $48400. It is understood that disparity in productivity will always exist due to differences in culture, work ethics, and political systems, but new partnership programs are imperative in order to encourage global benefits and discourage global instability.

[*] Prof. A. Zewail
Arthur Amos Noyes Laboratory of Chemical Physics
California Institute of Technology
Pasadena, CA 91125 (USA)
E-mail: zewail@caltech.edu
Prof. M. Zewail
Department of Chemistry and Biochemistry
Southwestern University, Georgetown, TX 78626 (USA)

For developing nations, the barriers to achieving developed-world status are many. High rates of illiteracy in some developing countries reflect the failure of educational systems and are linked to the alarming increase of unemployment. Second, the limited and ineffective use of human capital—largely due to strong seniority-based systems, nepotism, and the centralization of power that thwart opportunities for advancement—results in suppressed thought and stifles human potential. Third, the intermingling of state laws and religious beliefs causes confusion and chaos through the misuse in politics of the fundamental religious message of ethics and morality. And fourth, there is no coherent vision for education and science policies.

The developed world is now facing problems of similar magnitude but the problems are different in origins and solutions. One critical problem is the change in the state of education and scientific research. After the Second World War, scientific research was well supported: in the 1960s, for example, the sky was the limit, certainly in the U.S. One of the authors (A.Z.) benefited greatly from the conducive atmosphere for curiosity-driven research, and during this time his research at Caltech began with a funded proposal submitted to the U.S. National Science Foundation. The research described there was esoteric by today's standards, proposing to examine the coherence of atoms and molecules—a subject that was to be explored out of sheer curiosity. The truth is that the investigator did not plan for this research with the broad impact on society in mind; that is impossible to foresee. This research was the foundation for the work that led to the 1999 Nobel Prize in Chemistry for the development of *femtochemistry*. It is doubtful that this same research proposal would be funded today.

At present, the situation is different as a result of major changes in the funding policy of basic research and in the principles governing the quest for new knowledge. Because resources are limited, universities and research institutions have been forced to reorient their missions more towards profit making. Surely, the ultimate benefit of useful knowledge is to serve society's needs, but as history tells us most paradigm-shifting innovations come from researchers' curiosity about natural phenomena, and in many cases scientific discoveries have been made through serendipity. Even the most distinguished researchers cannot predict the path to the next discovery or the most important innovation. Unpredictability is the fabric of scientific discovery.[2]

We are both connected to Caltech academically and have observed at least two generations of researchers and students.

It is clear that the majority of those who strive for an academic career are driven by the excitement of research. These aspirations are then fostered by the values and knowledge derived through a good education system and by government and society's appreciation of the importance of education and science. In general, young people choose science and engineering as a profession, in academia or industry, because they can acquire new skills and learn the rational scientific thinking required for solving problems. In return, they expect a decent job.

Today, finding employment in, for example, chemistry, has become challenging. In fact, according to the most recent American Chemical Society survey of new graduates in the U.S., unemployment in chemistry reached a record high. Due to economic forces and recent trends in the pharmaceutical industry, many PhD chemists are holding temporary or non-chemistry-related positions or are unemployed. In addition, the average age at which principle investigators in the United States receive their first Research Project Grants (R01 awards) from the U.S. National Institutes of Health has increased to 42 years.[3] This is related, at least in part, to the increased average age at which investigators obtain their first academic appointment, and multiple postdoctoral positions are often becoming necessary. Needless to say, these obstacles and postponements may discourage young people from pursuing an independent career in a research environment.

Industrial research labs have, for the most part, redirected their resources and efforts into applied research areas relevant to their market products. One jewel from earlier times was Bell Labs. At Bell Labs, fundamental research was so advanced that it used to be said it was "the best university in America." Numerous scientific discoveries and inventions were made there, including the transistor, which is the most basic unit that triggered the digital revolution. Bell Labs employed some of the world's best scientists and engineers and collectively they have had stellar impact on discovery and invention. The list included Nobel laureates Charles Townes, William Shockley, John Bardeen, Arno Penzias, and Robert Wilson, and the topics varied from masers and lasers to superconductivity and cosmology. Sadly, this structure for broad-based, curiosity-driven research at Bell Labs no longer exists.

Even in academia, curiosity-driven research is often looked upon unfavorably. Research proposals have to address "broad relevance to society" and provide "transformative solutions" even before research begins. In today's proposals, the number of pages dedicated to funds management, broad impact, and societal implications is significant, and in some cases dominant, when compared to the number of pages focusing on the science itself. Universities are increasingly pressured to raise funds for annual costs and overhead is on the rise. Professors are chasing funds and writing more proposals; the available time for creative thinking is reduced and the numbers of academics involved in commercial enterprise is increasing. Faculty tenure at universities is often determined by how much money young faculty members have managed to raise. These constraints and practices beg the question: Would a young Maxwell, Pauling, or Crick today be attracted to their professions, and would they be able to

pursue their inquiries into fundamental questions in the current environment?

For centuries, curiosity in science has led to discoveries and innovations, and this "old-fashioned" approach has had a proven impact on science and society. Quantum mechanics, relativity, the Big Bang Theory and the origin of the universe, and the deciphering of the genetic code are discoveries made along this path, and so are the inventions of revolutionary technologies such as the laser (conceived by asking how light can be amplified), MRI (developed from curiosity-driven research about spinning electrons and nuclei), and the transistor (discovered as a result of curiosity about the nature of electrons in semiconductors), not to mention the huge progress made in the chemical industry, the pharmaceutical enterprise, and the agricultural sector. The manufacturing, medical, and digital IT industries that followed now constitute the backbone of global communications and the economy. From the cell phone to Google's and Apple's global markets, basic knowledge derived from curiosity-driven research is the springboard of their developments.

What is clear is that progress in research requires the nurturing of creative scientists in an environment that encourages interactions between researchers and collaborations across different fields. But such a nurturing atmosphere cannot and should not be orchestrated by weighty management, as creative minds and bureaucracies are inharmonious. Today, officials in many developing countries are seeking mechanisms to reach the innovation level of the developed world, but the core principles of innovation are often misunderstood. Regrettably, the same trend is creeping into developed countries.

One must then ask, is there a formula for "managing discovery making?".[4] The answer is in the realization of and belief in the natural evolution of developments, from basic research to technology transfer, and then to societal benefits. For basic, fundamental research to flourish, three elements cannot be forgotten. First, and most important, are the people involved. A proper education in science, technology, engineering, and math (STEM) is essential, and the search for the most creative minds for carrying out research is the key for success in R&D. Large buildings, and even massive funding, alone will not lead to significant progress without such people.

Second, nurturing an atmosphere of intellectual exchange is of paramount importance for ideas to crystallize. The distraction for faculty of either writing extensive and numerous proposals or becoming business-like managers is the beginning of the end. There are serendipitous innovations that are made by a single person, or a group of researchers, but most discoveries and revolutionary ideas are a result of individuals having the time to think and interact with other researchers. Consider how quantum mechanics, a major force behind innovations in the world economy, came about over two decades of work and by great curiosity-driven thinkers, such as Planck, Bohr, Einstein, de Broglie, Schrodinger, and Heisenberg.

Third, even the most creative mind can achieve little without resources. Obviously, investment in science is required to acquire new instruments and to hire competent staff. Countries and institutions that provide the proper

Angewandte
International Edition *Chemie*

infrastructure and the funding for new ideas will be the home of new discoveries. However, such support should follow the vision of creative researchers, rather than be based on the development of centers built to lure money or follow research fads. There are several examples of institutions of discovery in the developed world and in every case we see a correlation between the quality of the researchers involved and the significant discoveries made. Think of the Cavendish Laboratory in Cambridge under the leadership of J. C. Maxwell, J. J. Thomson, and L. Bragg and the discoveries that resulted there and changed the world: the electromagnetic nature of light, the discovery of the electron, and the development of X-ray diffraction. Caltech is another example; we find it remarkable that an institution with less than 300 faculty members in all disciplines is able to produce from its faculty and graduates 35 Nobel laureates. The key to these achievements is the R&D milieu that has flourished there from its inception, and that still attracts the best minds.

The correlation between investment in basic research and innovations that impact economies and societies is clear across the globe, from the U.S. after World War II to China today. In the 1950s, Robert Solow (1987 Nobel Prize in Economics) showed that new technologies create a large portion of economic growth, affecting nearly 75 % of the U.S. growth output. The theory of quantum mechanics alone continues to have major impact on world economy as without it revolutionary technologies would not have been realized. Think of the laser and the optical communication industry, MRI and the health industry, and the transistor and the IT market, not to mention the vast progress made in drug discovery, gene technology, and miniaturization. Vannevar Bush's vision of science as the endless frontier[5] is indeed the one vision legislators should keep in mind when deciding on funding policy of basic research, innovations of tomorrow, and economic prosperity.

Sunday August 5, 2012 marked the historic rover landing on Mars; this triumph demonstrated the accumulated power of innovation in the U.S. It is fitting that the one-ton, car-like rover begat by curiosity bears its name—*Curiosity*.[6] Over the coming two years it will reveal unknowns of our nearby planet. This quest for knowledge of the unknown is what science is all about, and is how the developed world acquired its power of innovation. Some developing countries are realizing this power of science, investing generously to transform their economies, and the progress is evident. Equally important is their investment in science education. At the 2012 International Chemistry Olympiad, the team from South Korea emerged as the top performer among 72 countries in the event held in the U.S.; according to C&EN magazine, "*members of the U.S. put on a solid performance*".[7]

Since the Industrial Revolution, the West has dominated world politics and economics principally due to the power of science, but it would be hubristic and naive to think that we now know what is relevant for tomorrow. Investing in education is investing in the future. For the past five centuries the West has had the right ingredients for achieving progress through such investments in useful knowledge. Now it is time to return to this wise vision, perhaps with a new intra-national partnership between government, industry, and research

institutions. If not, a transition may be in the making, with the sun of innovation rising from the East.[6]

Epilogue: The Genius of Science

Everyone knows the value of the Nobel Prize, but perhaps many do not know what is engraved on the other side of the medal bearing the famous portrait of Mr. Alfred Nobel. The medal was designed by Erik Lindberg in 1902 to represent Nature in the form of a goddess resembling Isis—or eesis—the Egyptian Goddess of Motherhood, Nature, and Magic (Figure 1). She emerges from the clouds, holding a cornucopia

Figure 1. Reverse side of the Nobel Prize medal.

in her arms and the veil which covers her cold and austere face is held up by the Genius of Science.[8] The inscription reads: "*Inventas vitam juvat excoluisse per artes*" (from Vergil, loosely translated as "And they who bettered life on earth by their newly found mastery"), with the official translation here "*Inventions enhance life which is beautified through art*". Indeed it is the genius of science which dispels the darkness of ignorance and lightens human lives for generations to come.

Received: August 20, 2012
Published online: December 3, 2012

[1] "Science for the Have-nots", A. H. Zewail, *Nature* **2001**, *410*, 741.
[2] "Unpredictability and Chance in Scientific Progress" J. M. Thomas, *Progress in Informatics* **2007**, 1.
[3] http://grants.nih.gov/grants/new_investigators/#data.
[4] "Curiouser and Curiouser: Managing Discovery Making", A. H. Zewail, *Nature* **2010**, *468*, 347.
[5] Vannevar Bush, "Science—The Endless Frontier". A Report by Bush, Director of the Office of Scientific Research and Development, to the President on a Program for Postwar Scientific

Research, National Science Foundation: **1960** (reprint); first published in **1945**.

[6] "How Curiosity Begat Curiosity", A. H. Zewail, *Los Angeles Times*, Sunday August 19, **2012**.

[7] "South Korea Tops Chemistry Olympiad", *Chem. Eng. News*, August 6, **2012**, p. 8.

[8] A. H. Zewail, *Les Prix Nobel, The Nobel Prizes 1999: Nobel Prizes* (Ed.: T. Frängsmyr), Almqvist & Wiksell, Stockholm, **2000**, pp. 103–203.

50 years from today

60 of the World's Greatest Minds
Share Their Vision of the Next Half Century

ASTROPHYSICISTS • INTERNET PIONEERS • CONSERVATION BIOLOGISTS • NEUROLOGISTS • MILITARY COMMANDERS
WORLD LEADERS • ECONOMISTS • INVENTORS • NATIONAL MEDAL OF TECHNOLOGY WINNERS • ASTRONOMERS

EDITED BY

MIKE WALLACE

Ahmed Zewail

Ahmed Zewail was awarded the 1999 Nobel Prize in Chemistry for his pioneering development in femtoscience, making possible the observation of phenomena in a millionth of a billionth of a second. He is chair professor of chemistry and physics and director of the Center for Physical Biology at Caltech. Postage stamps have been issued to honor his contributions to science and humanity.

THE WORLD IN FIFTY YEARS: REVOLUTIONS AND REPERCUSSIONS

Traveling through time is a real challenge. Throughout history, predictions of the future have been claimed by notables and later proved to be contrary to reality. We usually think in the frame of reference of the "present" without, for most people, being able to see into the "future"—crystal balls do not usually work. The founder of IBM, Thomas J. Watson, about the future of computers said in 1943, "I think there is a world market for about five computers." Multiply this number by a billion to reach the reality of our time! Nevertheless, it is interesting to conjecture the state of the twenty-first century because of its uniqueness in the history of human knowledge. In the coming fifty years, any analysis of the state of progress must consider forces of knowledge and faith, affected, of course, by the political, economic, and natural changes of influence.

There is no doubt that one can predict, based on current trends, that science and technology will provide new discoveries and innovations in many fields of endeavor. Here, I first highlight advances that will span vast scales of length and time—from the micro, very small (atomic scale), to the macro, very big (cosmos scale), and to the very complex (life), the scale in between. For the micro world, it will become possible to tame and manipulate matter on the nanometer (a billionth of a meter) length scale and on the femtosecond to attosecond (a millionth of a billionth, and a billionth of a billionth, respectively) timescale. This control with precision may lead to the synthesis of robotic micro-machines that can form "intelligent matter," matter with a specific molecular-scale function, or the building of a biological-mimic factory, the cell. Drug design from first principles would have a huge impact on the cure of diseases, hopefully at much reduced prices for the needy, especially those in less developed countries. Academically, such multidisciplinary approaches, requiring knowledge of various physical and biological sciences, will restructure university education and redefine new fields of study at the interface.

Building on advances in molecular and cellular developments, humankind will reach a far better understanding of the biological function of organs, such as the brain, but I doubt that the code of consciousness will be cracked even in fifty years of research. New tools will be developed to observe the behavior of complex systems in space and time, and new concepts describing complexity may emerge. Treatment of diseases such as Alzheimer's will take a new course— control at the level of the molecule (protein); the targeting of genes (DNA); or the making of spare parts organs using stem cells and cloning techniques aided by current methods of genetic engineering, molecular biology, and gene expression (PCR, RNAi, recombinant DNA . . .). In the coming fifty years, life spans may extend beyond one hundred years, and human health care will shift more from the office of physicians toward homes in what I call "PM care," or personal medicine. But medical and societal benefits will not evolve without some repercussions. The synthesis or control of physical matter is vastly different from the interference with or modification of life (e.g., using silicon instead of carbon as a base), and to many on this planet this intrusion represents a conflict with religious belief. Without improved education and open discourse on ethics and morality, clashes of science and society may become of serious consequence, even in developed countries such as the United States.

Progress will not only be made in the world of matter and life, but also in the cosmos at large. More than 80 percent of our universe is made of dark energy, and we at present do not really understand its nature. The physical observed matter is only about 5 percent and the rest is dark matter—matter of unknown composition, which cannot be observed optically, but whose presence is inferred from the effect of gravity on visible matter. In this century, cosmology as a science will provide a better understanding of dark energy and matter. We may also uncover some mysteries of black holes by detecting gravity waves, hopefully with new surprises beyond the predictions of Einstein's general theory of relativity. Space missions will continue to explore the solar system and beyond, and we may learn more about the involvement of planets in the origin-of-life question. Colonies may be formed on the Moon—hopefully not for military purposes— and commercial spaceship travel may become the new Dreamland, the Disneyland that people will enjoy trekking to, but at a price!

In technology, it is safe to say that the most rapidly changing development will be that of information and computers. The leaps forward will result from the reduction of the physical size of computers and the increase in their computation capacity and speed, possibly beyond the limit of (Gordon) Moore's Law—an empirical observation in 1965 that held that the number of transistors on a single chip (integrated circuit) will double nearly every two years. The expansion of the wideband network and the integration of multimedia into one device, not to mention the possibility of quantum computing using the language of the atom (quantum mechanics), will have a major impact on information communication, speed and capacity, and on society and culture. Society will readjust to the information explosion, but we should expect that traditional family values may be further compromised.

Unfortunately, despite these anticipated discoveries and innovations, there are daunting chasms to cross in the coming fifty years. Advanced, and even simple, technologies will empower some people to inflict large-scale disasters through "terrorist acts" or "freedom-fighting causes." Given the 80 percent population of the have-nots on our planet, a large fraction of people are desperate or frustrated because of their poor economic or unfavorable political situation, and they will somehow blow off their steam and unleash significant damages in large-scale populations. Chemical, biological, or nuclear weapons, such as dirty bombs, may be used, and, as a result, the response may trigger the first big nuclear war between nations. It would be wise to alleviate, or at least ease, the despair of the have-nots, and to sustain economic and education aid guided by bridging dialogues, especially between the West and the 1.3 billion Muslims. Judging from the performance of today's world leaders, and despite human progress, I fear the continuing dearth of visionary solutions to the underlying causes of human misery—ignorance and deprivation.

In the coming fifty years, the only superpower of today's world may have real competition, and the challenges are both international and national. In an interdependent world, the United States can no longer afford to have a fragmented and incoherent foreign policy. As other nations build up to a leading economic position, the United States cannot allow the erosion of innovation and investment in R/D, decline in work ethics, and the weakening of the traditionally powerful middle class. As importantly, the unique value system that guided the nation and attracted the best minds from around the world cannot be convoluted by the political gains of elected officials and lobbyists. All of these factors will determine the global position of the United States in fifty years. Already economic indicators forecast that by 2050, China will have a projected GDP of $45 trillion, surpassing the United States, with India's GDP approaching that of the United States, and the dollar is losing its prime position among all currencies. I, however, still believe in the American

Despite human progress, I fear the continuing dearth of visionary solutions to the underlying causes of human misery—ignorance and deprivation.

system for creativity, and hopefully in the coming fifty years the country will restructure its political and education systems to preserve the values that defined its uniqueness in history.

The above exposé into science, technology, and society does not take into account natural forces that bring about disasters. These include consequences of climate change, major epidemics or earthquakes, hurricanes such as Katrina, and possibly a 50–100 km asteroid impacting the earth in an unprecedented collision. Surprises such as 9/11, Hiroshima's mass destruction, or a viral/mechanical breakdown of e-communication may occur before fifty years. But one major force beyond nature that is surely here to stay is religion. It would be naïve to ignore the importance of faith in the lives of billions of people all over the globe. In fifty years' time, many of the troubles that currently plague the world can be considerably mitigated if, in the vision of policy and diplomacy, we accept reason and faith, which need not be in conflict, as the dual human need for knowledge and meaning of life on this planet.

In closing, I believe that the coming fifty years will bring about revolutionary discoveries and innovations as humankind acquires new knowledge. Humans, however, are still the same species—Homo sapiens—that desire the use of force in all forms, as primitive as fire and as sophisticated as nuclear bombs. What is different in the twenty-first century is the ability of a few individuals to cause massive destruction. It is clear, at least to me, that in the coming fifty years and beyond, the picture is, on one hand, rosy for the scientific and medical advances that will positively transform lives, and, on the other hand, gloomy when considering the lack of vision in global affairs. Progress can only be realized when political leaders see the world through the lens of human rights and peaceful coexistence. Only then can the new generations fully benefit from the tremendous potential of the twenty-first century.

R. Braun and D. Krieger (Eds.)

Einstein – Peace Now!

Visions and Ideas

⊗WILEY-VCH

The Future of Our World[1]

Ahmed Zewail

Over the last century, our world has experienced at times a "beautiful age" with promises of peace and prosperity, but then some imposing forces changed the entire landscape. History reminds us of recurrences, and the current state of the world is not so different that we may ask – what political and economic forces cause such disorder in a world seeking prosperity through globalization and revolutionary advances in technology? Here we will address the need for a rational world vision that must take into account developments in the population of the have-nots and dialogues of cultures. It is a vision of economic, political, religious, and cultural dimensions in world affairs. Only with such a vision can we shape a bright future for our world.

Excellencies, Ladies and Gentlemen

It is a great honor to give this year's U Thant Distinguished Lecture at the United Nations University in Tokyo. I applaud the purpose of the lectureship named in honor of Mr. U Thant, the Secretary-General of the United Nations from 1961 to 1971 and the man who had the vision to establish this University. I would like to take this opportunity to thank the UNU Rector, Professor van Ginkel, the Director of the Institute of Advanced Studies, Professor Zakri, and the President of the Science Council of Japan, Professor Yoshikawa, for making this event possible with thoughtfulness and style. I also wish

1) 5th U Thant Distinguished Lecture Series, United Nations University, Tokyo, April 15, 2003. 1st Lecture given by Prime Minister of Malaysia, Mahathir Mohamad; 2nd by President of the Republic of South Africa, Thabo Mbeki; 3rd by President of the United States, William J. Clinton; 4th by Nobel Peace Prize Winner, Norman E. Borlaug.

to acknowledge the Ambassador of Egypt Dr. Mahmoud Karem for his warm welcome.

Last year the lecturer in this series, President Bill Clinton, spoke about globalization and our shared future, and the year before Prime Minister Dr. Mahathir Mohamad spoke about globalization, global community, and the UN. Both speakers were concerned about the new emerging world and opportunities for prosperity and global unity. Today I would like to share with you my thoughts about the future of a turbulent world in the present state of economic and political disorder.

The title of my lecture has several implications that I should clarify. It gives the impression that I know the future or the science of futurology. I do not, and in fact I am aware of many predictions made in the past that turned out to be incorrect. What I have in mind is to paint a future that benefits from our history and our rational thinking; a future shaped by *Homo Sapiens* – the species with the greatest brain power on Earth. For this future, I shall present what I envisage for a world of peace and prosperity and how we can achieve our goals with justice and fairness. But first, let me take you inside a time machine to "travel in time" and see what history will tell us.

The Beautiful Age

Nearly a century ago, the world of 1870 to 1914 had an optimistic outlook. The French called the decades before WWI, which broke out in 1914, "The Beautiful Age – La Belle Époque". The world was experiencing the same upbeat spirit of the global community that Mr. Clinton and Dr. Mohamad spoke about here in our present world. Peace and prosperity were on the horizon. The material standard of living was on the rise, democratization was on the rise, continents were being connected by railroads, steam ships, automobiles, airplanes, the telegraph, and the telephone. Man conquered the last uncharted territory of world maps, the North Pole in 1909, the South Pole in 1911, and the United States became the land of promise for millions. Achievements in the sciences, literature, and peace were honored by the awarding of the first Nobel Prize in 1901. (The Nobel Peace Prize centennial publication has indeed given the reasons for calling this period the Beautiful Age.) At that time, the principal force was the force of science and technology that was creating a better life for humankind.

What went wrong then? Great powers were hungry to conquer lands and resources in Africa, Asia, and the Pacific. Control over raw materials and markets and strategic positioning in the world were the driving forces. The power gained by industrialized nations gave them a thirst for the right to rule, and in some cases oppress, those who did not have power. People from other parts of the world could only acquire a Western level of advancement by learning to think like Westerners, and missionaries often defined civilization as a combination of Western religion and science.

The Great Powers formed alliances and Europe was experiencing nationalism. Germany, Austria–Hungary and Italy formed the Triple Alliance, and France, Russia and Britain formed the Triple Entente. The empires of Russia and Austria-Hungary competed for influence over the Balkans following the disintegration of the Ottoman Empire, "Europe's dying man". The First World War began, and the rest is history.

The World of Today

Today, one hundred years later, the analogy may be telling of the dynamics in our present world. In the recent era of globalization (1991–2000), the world looked beautiful again, coming together by the force of global economy and global political ties. The policy of apartheid in South Africa was abandoned, Nelson Mandela was released from prison, and was elected President in 1994. Even the Gulf War of 1991 – strategic to the control of oil resources – appeared to have a moral dimension, namely the return of Kuwait to its people. The solution to the Palestinian–Israeli conflict was on a hopeful track with the signing of the Oslo Accord in 1993. European economic and political cooperation took on a new dimension with the creation of the European Union, and Japan and other so-called Asian Tigers took a major role in world economic developments. United Germany gave the world a new hope for unity and the end of an era – the world of 1946–1963. This world of the Cold War and nuclear armament appeared to have changed into a world of globalization in the 1990s.

Science and technology were again the real forces for achieving the new world status. Information technology brought the world to village-type communities. Advances in the new knowledge of lasers, semiconductors, biotechnology, and the like have transformed our lives with revolutionary improvements in communication and health, and we even began to dream about a future on other planets.

This is not to say that the world now is perfect. Conflicts are still raging in parts of Africa and HIV/AIDS continues to take the lives of large numbers of people in the sub-Saharan countries. Human rights violations and occupation by force continue in the world of globalization. As I speak today, the Iraqi war is taking the lives of innocent people, and the Palestinians are still under occupation. In Europe, there was the horrific ethnic cleansing in the Balkans and the conflict between the Catholics and Protestants in Northern Ireland continues to this day.

Notwithstanding these conflicts and disorders, the nations of the world on the whole are aiming for a united globe through understanding and cooperation – the role of the UN – and through economic developments – the role of globalization. The desire to achieve more peace and stability through global cooperation is articulated, for example, in the Millennium Development Goals (MDGS), a decree endorsed by all member states of the United Nations in September 2000 with the objectives of attacking global problems such as poverty, diseases, and education for all people, from Nairobi to New York. Through cooperation, many agreements and accords have been reached: the disarmament agreement between the United States and Russia known as START (Strategic Arms Reduction Treaty); the peace agreement for a NATO and Russian partnership; the agreement for the banning of landmines; the UN International War Crimes Tribunal; and global conferences to address problems such as the environment, water resources, and AIDS.

World Disorder and Superpower

What then is causing the current disorder? In my view there is a short-term cause and a long-term problem. The September 11, 2001 horrific attack on the United States has caused an impulsive impact on the only superpower in the world. The country has been insulated from external wars throughout its history, geographically distanced by the Atlantic and Pacific Oceans. Moreover, the political system, which is greatly influenced by strong lobbying and capitalistic media, at times has created a gulf between the United States and other countries. The United States is a unique country and the diversity in its population has resulted in an amalgamated culture – a melting pot. But this new culture is not necessarily knowledgeable about the original cultures of its people. The United States also knows it is especially unique and possesses the ultimate power – that of science and technology – and this power makes it a ruling force over world economy, markets, and military status.

America is still in a state of shock and disbelief, and the response in the country varies from moderate to fanatic. Sadly, the September 11th attack may have coincided with extreme religious and political agendas. With this confusion and confluence, now is the time that America needs visionary leadership the most. A vision of unity is in the best interests of all. The United States has a responsibility to lead the globe into becoming a united world. The world still remembers the Marshall Plan and the Peace Corps as examples of uniting initiatives. America cannot afford to alienate people around the world, and it must apply the same standards of fairness at home and abroad. We must all look for the real sources that kindle terrorism and not try to camouflage the real reasons behind it. In the long run, the key is not to ignore the have-nots, not to ignore the frustrated part of the world, politically and economically, and to recognize that poverty and hopelessness are the primary sources of terrorism and the disruption of world order.

The World of the Have-Nots

In our present world, the distribution of wealth is skewed. Only 20% of the population enjoy the benefit of life in the developed world, and the gap between the haves and have-nots continues to increase. According to the World Bank, out of the 6 billion people on Earth, 4.8 billion live in developing countries, 3 billion live on less than $2 a day, and 1.2 billion live on less than $1 a day – an amount that defines the absolute poverty standard. About 1.5 billion people have no access to clean water, with health consequences like waterborne disease, and about 2 billion people are still waiting to benefit from the power of the industrial revolution. The per capita gross domestic product (GDP) in some developed Western countries is $35,000, compared with about $1,000 per year in many developing countries, and significantly less in underdeveloped countries.

This difference in living standards by a factor of 100 or so between the haves and have-nots ultimately creates dissatisfaction, violence, and racial and ethnic conflict. Evidence of such dissatisfaction already exists and we have only to look at the borders of developed and developing or underdeveloped countries; for example, between the United States and Mexico and between Eastern and Western Europe, or between the rich and the poor in any nation. Of similar effect is the frustration caused by double standards in international disputes and in the support of undemocratic or even corrupt regimes for the sake of national economic or political gains.

Some believe that globalization is the solution to problems such as the economic gap, the population explosion, and social disorder. Globalization, in principle, is a hopeful ideal that aspires to help nations prosper and advance through participation in the world market. In practice, however, globalization is better tailored to the prospects of the able and the strong, and, although of value to human competition and progress, it serves only that fraction of the world's population who are able to exploit the market and the available resources.

Nations must be ready to enter through this gate of globalization and such entry has its requirements. Among these requirements are the following: computer and internet literacy, a minimal level of bureaucracy, accessibility to sources of knowledge and information, the entrepreneurial spirit, efficiency in management, and the clear and just applications of the law. With a new system of education and the development of a science base we can hope for an effective globalization. These cannot be achieved without partnership.

World Partnership and World Science

It is clear that world order requires a new and comprehensive partnership between the developed and developing worlds. In my view, science and education are the real glue for binding different cultures and for achieving progress and prosperity. Science is the only fundamental and international language of the world. The developed world is developed because of its scientific and technological power. In this century, knowledge-based societies will capture the lion's share of the world economy and prestige. But, how can the developing world reach such a high state of achievement in science and use the benefits for the betterment of the have-nots? And what will the benefits be to the developed countries?

In the past five years, the scientific community worldwide has published about 3.5 million research papers. Europe's share is 37 per cent. The share of the United States is 34 percent. The Asia/Pacific share is 22 per cent. Other places, representing 70 to 80 per cent of the world's population living largely in the developing world, contributed less than 7 per cent of these scientific articles. Put in a different way, as Mr. Kofi Annan pointed out recently, "Ninety-five per cent of the new science in the world is created in countries comprising only one-fifth of the world's population. And much of that science – in the realm of health, for example – neglects the problems that afflict most of the world's people."

What difference does this disparity in academic output make? Should only universities and research centers be concerned? I do not think so. Consider this interesting correlation. The United States' contribution to the world's annual economic output is between 30 and 40 percent, comparable to its share of scientific output on a global scale. Europe's annual economic output registers a similar percentage and, like the United States, its economic output tracks its contribution to its output of scientific and technological contributions. It is unlikely that this correlation is coincidental.

If we are aware of these trends and understand the problems that stand in the way of progress, why do we have such difficulties building scientific capacity in the developing world and putting science to work to improve its economic well-being? First, the developing world must get its house in order. A renaissance in thinking is needed – we need to pay more attention to education and we should invest more in science and technology. The objective is to provide a new work force equipped with 21st century tools of education and skills and with a belief in ethics and team work. We also need to lower the bureaucratic political barriers that stand in the way of success and to rule by laws that allow for the freedom of thought. Women must participate as full partners in our pursuit of knowledge.

The developing world possesses very capable scientists and continues to contribute outstanding scientists unwittingly to the developed world as part of the brain-drain phenomenon. At Caltech, for instance, my own research group is more than 50% Asian. But major reforms of the system are badly needed. Clearly, this may not be possible on a grand scale in a short time, but the foundation must be established properly and in a timely manner. Empty slogans, or waiting for the developed world to solve the problems, or blaming people of the developed world with conspiracy theories will not provide the means. Yes, international politics play a role, but people's will is a stronger force, provided the force is coherent and not dispersed by internal politics.

The developed world must also carry important responsibilities for its share in partnership, in building the scientific and human capacity in the developing world. First and foremost it must reform its international aid programs, investing less money on military hardware and instruction and more on scientific training and partnerships. Some of the money spent on fighter planes, not to mention the recent war, could fund research programs all over developing countries, helping in what must be the ultimate goal – global education and prosperity through developments. Politics, moreover, should be drained from international aid programs to ensure money is available for productive initiatives that could help boost science and technology in the developing world.

What will rich countries receive in return for the help they give the have-nots? First, there is the moral dimension. The psychological value derived from being a generous global neighbor should not be underestimated. Even on a personal level, most of us do try to help each other, and all major religions encourage and legitimize helping the needy. It is also important to realize that the prosperity of the developed world is in part due to natural and human resources from the developing world, and their markets.

Second, the developed world should acknowledge the importance of reciprocation over time. Islamic civilization gave a great deal to Europe, especially during the Dark Ages. The Arab and Islamic civilizations were major contributors to the European Renaissance. The Islamic civilization rose to become the foremost economic power in the world, at which time it also reached the highest level in the sciences. Today the Muslim world needs help and there is nothing wrong with the United States, Europe, Japan, and other developed nations lending a hand as a modest gesture to the changing fortunes of history.

Third, there is a practical, self-centered consideration based on the time-tested importance of having an adequate insurance policy. In the United States, I pay a great deal for insurance to protect my family against the high cost of medical care, to protect our house against fire and theft, and to protect our cars against accidents. Similarly, the developed world needs to invest in an insurance policy to help it live in a safer and more secure world, but it better be a genuine and good policy!

The choice for the haves is clear. They have to be involved. The choice for the have-nots is also clear. They have to first get their house in order, and build the confidence for a transition to a developed-world status. The transition is possible. In a meeting with Prime Minister Mahathir Mohamad on a recent visit to Malaysia, I learned of the critical role of the new education system implemented during the nation's rapid transition from a labor-intensive economy dependent on cheap labor to a knowledge-based economy poised on the doorstep of the developed world. It is a transition that has been fueled by the belief in building the proper base for modern technology.

The 21st Century – Future Frontiers

Technology in the 21st century is knowledge-based, and unskilled cheap labor, which may have worked for developing countries in the past, will not work in this century. How can the developing world embrace economy-transforming technologies like microcomputing, genetic engineering and biotechnology, information technologies, and femto and nano-technologies without a strong foundation in science? Does the developing world always have to wait decades before participating in global science and technology? Can nations become a part of the modern world without losing their cultural and religious identities? The new century promises unlimited opportunities in science and technology and I believe that the developing world can and should be a partner in or a part of that development. I would like to mention today the new frontiers encompassing three scales in our universe and which I recently outlined:

Our matter – the scale of the very small. We are on our way to being able to manipulate matter at its smallest, most fundamental limits both in time, on the femtosecond scale, and in length, on the nanoscale. Just think about these new scales of time and space in the world of the very small. If your heart beats once a second, now we can see the beats of atoms in a femtosecond, in a millionth of a billionth of a second – a femtosecond is to a minute as a minute is to the age of the universe. Similarly, we can study matter on the nanometer scale and resolve the atoms in their structures – the size of the atom to the size of the earth is like the size of the earth to the whole universe. The opportunity is huge for acquiring new knowledge and for creating new forms of "our matter". The manipulation of matter to produce new sources of energy (photovoltaic/photosynthetic, etc.) should become a major undertaking. The interface of matter's micro- and nanonetworks, designed to produce artificial intelligence, to our life organs, such as the brain, will be another frontier that could alter the boundaries and meaning of species.

Our universe – the scale of the very big. In this century, we may have colonies on the moon, and we may have our second homes on other planets and maybe even in other solar systems. Just think of the scales of the world of the very big. Our universe is about 12 billion years old, and at the speed of light (300,000 km/s), our universe's limit of distance is 100 billion trillion kilometers – certainly enough space for the six billion people on earth today, even multiplying by ten or by one million in the future! The opportunities involving outer space and information technology are unlimited. On our planet, any information one needs will be provided through the so-called "virtual walls" and education and intelligence in all societies will have to be redefined.

Our life – the scale in between. In the first year of this century, the sequencing of the human genome was completed. We now have the genetic map that describes every human on planet Earth. Just think, three billion letters have been deciphered and read into our book of life. The history of biology has changed from the classification of living organisms (Darwin's theory), to the world of cells (Leeuwenhoek–Hooke microscope), to now the molecular world (Watson and

Crick's DNA) with revolutionary ideas in genetic engineering and biotechnology. Soon we might see a nanoscale motor entering the cell to do work. Medicine and human health will certainly enter a new age.

It seems to me that indeed opportunities are unlimited and with education, skill, and the brain power of people around the globe, we should be able to benefit and share in the wealth of new discoveries and technological advances. But, we must first learn how to live together on one globe.

Confluence of Civilizations

The developed and developing nations, aside from their economic and political ties, have to participate in a dialogue among civilizations and a dialogue among cultures. Some intellectuals have introduced concepts such as the "clash of civilizations," as termed by Samuel Huntington, and the "end of history," as expressed by Francis Fukuyama. Both authors argue their cases with conviction, nonetheless, these ideas are controversial and debatable.

As a scientist, I find no "fundamental physics" in these concepts. In other words, it is not a fundamental principle of civilizations that they be in a state of clash with each other. Neither is it a fundamental principle to end history with one system over all other ideologies. I argue that the current world disorder results in part from ignorance about civilizations – unawareness or selective memory of the past and lack of perspective for the future – and in part from the economic misery and political injustices (domestic and/or international) experienced by the have-nots.

According to the dictionary, civilization means an *advanced state* of human society in which a high level of culture, science, industry, and government has been reached. Individually, we are civilized when we reach the advanced state of being able to communicate with and respect those of different customs, cultures, and religions. Collectively, we speak of globalization as a means for bringing about prosperity in the world, yet globalization cannot be a practical concept if there are clashes of civilizations. Historically, there are many examples of civilizations that have coexisted without significant clashes.

I have written about these issues and it is perhaps useful to distill the main points here. The central argument in the thesis of the clash of civilization is that in this post – Cold War era, the most important distinctions among peoples are not ideological, political, or economic – they are cultural. Hence people define themselves in terms of ancestry, religion, language, history, values, customs, and institutions. According to this thesis the world becomes divided into eight major civilizations: Western, Orthodox, Chinese, Japanese, Muslim, Hindu, Latin American, and African.

I have several difficulties with this analysis, and perhaps the following questions and commentary may clarify my position. First, *what is the basis for these divisions of civilizations?* People belong to different cultures, nations have experienced (and continue to experience) different cultures, and nations on the same continent may be influenced by different civilizations. In my own case, from birth to the present time, I can identify myself as Egyptian, Arab, Muslim, African, Asian, Middle Eastern, Mediterranean, and American. Looking closely at just one of these civilizations, I note that the Egyptian people belong to a dynamic civilization with a multicultural heritage: Pharaonic, Coptic, Arabic, Islamic, not to mention the Persian, Hellenistic, Roman, and Ottoman influences.

Second, *is it fundamental that differences in cultures necessarily produce clashes?* In this thesis, it is contended that if the United States loses its European heritage (English language, Christian religion, and Protestant ethics), its future will be endangered. I reach the opposite conclusion. From a personal point of view, I did not speak English when I came to the United States, I am not a Christian, and I was not taught Protestant ethics. Yet I integrated into my new, American culture while preserving my native culture(s) and I believe that the confluence of my "Eastern" and "Western" cultures is without a clash. From a broader perspective, America's strength has traditionally risen from its "melting pot" culture; the country has been enriched and continues to be enriched by multiethnicity and the different cultures of its inhabitants.

Turning to international relations, cultures and civilizations can be at their peak of achievement and yet coexist in harmony and even complement each other. The United States, Japan, and European nations are examples of this beneficial coexistence created by building economic and cultural bridges. Another example comes from a country that many would have doubted had the potential for creating ethnic and religious harmony: Malaysia with its inhomogeneous population of Malays (53%), Chinese (26%), and Indians (8%) with different religions – Muslim (60%), Buddhist (19%), Christian (9%) and Hindu (6%). Neither religion nor culture seems to hinder its progress, and certainly Malaysia is an economic success story, demonstrating the civility of living harmoniously together with a variety of cultures.

Finally, *what about the dynamics of cultures?* Cultures are not static; they all change with time, and the degree of change is governed largely by forces of politics, economics, and religions. Let us consider my home country. Egypt's civilization was developed very early in human history and it dominated the world for millennia, but lately the nation has become a developing one. This does not mean that Egypt has lost its civilization, but it does mean that, like others, it has changed with time, due to many internal and external forces – the current state is not due to a genetic factor or fundamental cultural values. Cultural changes may impede progress within a nation, but not necessarily through clashes with others, or permanently.

What we have to consider seriously are the political and economic interactions within a culture and between the various cultures of the world. The people of North and South Korea are of similar culture, but the notable disparity in progress between the two countries is due to economic and political factors; the same can be said of East and West Germany before reunification. It is easy to divide the world into "us" and "them", and slogans such as the clash of civilizations or the conflict of religions certainly make it very difficult to unite the nations of the world – we need dialogues, not conflicts or clashes!

Epilogue

I would like to close with a message. The world at the beginning of the 21st century is divided, not only politically but also between hope and hopelessness. On one hand, progress can be seen in the human life – life expectancy has increased by ten years over the past three decades, infant mortality has fallen by 40%, and adult illiteracy has been reduced by half. On the other hand, every day 30,000 children die of preventable diseases, some 60 countries grow poorer, the spread of HIV/AIDS has become the most deadly epidemic in human history according to the UN, and the crisis of global water is becoming real as we witness in Iraq today and most probably in future conflicts on the whole planet. It is not naive to think of a better world and to achieve that goal through courage, justice, and liberty.

Judging from history, the shape of the future is made by leaders who have the capacity to turn it into an epoch of hope for peace and prosperity or into one of divisiveness and disorder. It is unlikely that those who do not know about history will make history. Leaders of the world should use the benefits of knowledge to shape a hopeful future for our children and grandchildren, for posterity. This can only be achieved by understanding the need for justice in the world and by promoting dialogues and cooperation among countries and peoples of the world, the world community. That is why it is vital to maintain a financially strong and independent UN. Despite the complexity in world affairs, undermining the UN as a viable institution for world education and peace will be a tragedy of enormous consequences. Even the superpower, the United States, whose population is only 5% of that of the total on planet Earth, cannot be the world's judge, jury, and executioner.

To shape our future in the age of globalization we need to develop a new perspective – one encompassing the economic, political, religious, and cultural dimensions of world affairs. Miss Kalpana Chawla, an Indian-born naturalized American who lost her life in the space shuttle Columbia disaster on February 1st of this year (2003) said, "When you are in space and look at the stars and the galaxy, you feel that you are not just from any particular piece of land, but from the solar system." She was viewing the world from the heavens and she had a universal perspective. A true statesman or stateswoman will see our world with a universal perspective that is unifying for humanity. And then wars may become wars on global poverty, disease, and despair, and for a sustainable world future.

References

[1] Øivind Stenersen, Ivar Libaek, Asle Sveen, *The Nobel Peace Prize*, Cappelen Forlag AS, Oslo (2001).

[2] The Economist, February 8, 2003 (for Ms. Kalpana Chawla); The World in 2003 (for Kofi Annan's MDGS); April 5, 2003 (for Malaysia's Report).

[3] Kofi Annan, Science, Volume 299, page 1485 (2003).

[4] Ahmed Zewail, Science for the Have-Nots, Nature, (London 2001), Vol. 410, p. 741.

[5] Ahmed Zewail, Dialogue of Civilizations, SSQ2/Journal, Routledge Press, Paris, France (2002); Address at UNESCO, April 20, 2002.

[6] Ahmed Zewail, Voyage through time – Walks of Life to the Nobel Prize, American University in Cairo Press (2002); Reprinted in 12 languages and editions.

Ahmed Zewail

Ahmed Zewail, was born in Egypt and studied at the University of Alexandria and University of Pennsylvania. He is presently the Linus Pauling Chair Professor of Chemistry and Professor of Physics, and the Director of the NSF Laboratory for Molecular Sciences (LMS) at Caltech, the California Institute of Technology, Pasadena, USA. Zewail has received numerous Prizes and Awards including the 1999 Nobel Prize in Chemistry for his pioneering developments in the field of femtoscience. Professor Zewail is also renowned for his tireless efforts to help the population of the have-nots.

The Abdus Salam
International Centre for Theoretical Physics

United Nations
Educational, Scientific
and Cultural Organization

International Atomic
Energy Agency

40th**anniversary**
1964
2004

ONE HUNDRED REASONS

TO BE A SCIENTIST

TRIESTE   2005   ITALY

IT IS POSSIBLE

Ahmed H. Zewail

California Institute of Technology, USA

© Courtesy of Ahmed H. Zewail

On the banks of the Nile, the Rosetta branch, I was born in Damanhur, the "City of Horus", only 60 km from Alexandria. In retrospect, it is remarkable that my childhood origins were flanked by two great places— Rosetta, the city where the famous Stone was discovered, and Alexandria, the home of ancient learning. I am the only son in a family of three sisters and two loving parents. My father was liked and respected by the city community—he was helpful, cheerful and very much enjoyed his life. He worked for the government and also had his own business. My mother, a good-natured, contented person, devoted all her life to her children and, in particular, to me. She was central to my life with her kindness, total devotion and native intelligence. Although our immediate family is small, the Zewails are well known in Damanhur.

The family's dream was to see me receive a high degree abroad and to return to become a university professor—on the door to my study room, a sign was placed reading, "Dr. Ahmed," even though I was still far from becoming a doctor. My father did live to see

that day, but a dear uncle did not. Uncle Rizk was special in my boyhood years and I learned much from him—an appreciation for critical analyses, an enjoyment of music, and of intermingling with the masses and intellectuals alike; he was respected for his wisdom, financially well-to-do, and self-educated. Culturally, my interests were focused—reading, music, some sports and playing backgammon. The great singer, Um Kulthum (actually named Kawkab Elsharq—a superstar of the East) had a major influence on my appreciation of music. Reading was and still is my real joy.

As a boy it was clear that my inclinations were toward the physical sciences. Mathematics, mechanics, and chemistry were among the fields that gave me a special satisfaction. Social sciences were not as attractive because in those days much emphasis was placed on memorization of subjects, names and the like, and for reasons unknown (to me), my mind kept asking "how" and "why". This characteristic has persisted from the beginning of my life. In my teens, I recall feeling a thrill when I solved a difficult problem in mechanics, for instance, considering all of the tricky operational forces of a car going uphill or downhill. Even though chemistry required some memorization, I was intrigued by the "mathematics of chemistry". It provides laboratory phenomena which, as a boy, I wanted to reproduce and understand. In my bedroom I constructed a small apparatus out of my mother's oil burner (for making Arabic coffee) and a few glass tubes, in order to see how wood is transformed into a burning gas and

Based on the biographical profile given in "*Les Prix Nobel: The Nobel Prizes 1999*" edited by Tore Frängsmyr, Almqvist & Wiksell, Stockholm, page 103 (2000).

a liquid substance. I still remember this vividly, not only for the science, but also for the danger of burning down our home! It is not clear why I developed this attraction to science at such an early stage.

After finishing high school, I applied to universities. In Egypt, you send your application to a central Bureau, and according to your grades, you are assigned a university, hopefully on your list of choices. In the sixties, Engineering, Medicine, Pharmacy, and Science were tops. I was admitted to Alexandria University and to the faculty of science. Here, luck played a crucial role because I had little to do with the decision, which gave me the career I still love most: science. At the time, I did not know the depth of this feeling, and, if accepted to another faculty, I probably would not have insisted on the faculty of science. The passion for science became apparent on the first day I went to the campus in Maharem Bek with my uncle—I had tears in my eyes as I felt the greatness of the university and the sacredness of its atmosphere. My grades throughout the next four years reflected this special passion. I graduated with the highest honors—"Distinction with First Class Honor". With these scores, I was awarded, as a student, a stipend every month of approximately £13, which was close to that of a university graduate who made £17 at the time!

After graduating with the degree of Bachelor of Science, I was appointed to a university position as a demonstrator ("Moeid"), to carry on research toward a Master's and then a Ph.D. degree, and to teach undergraduates at the University of Alexandria. This was a tenured position, guaranteeing a faculty appointment at the University. In teaching, I was successful to the point that, although not yet a professor, I gave "professorial lectures" to help students after the Professor had given his lecture. Through this experience I discovered an affinity and enjoyment of explaining science and natural phenomena in the clearest and simplest way. The students (500 or more) enriched this sense with the appreciation they expressed. At the age of 21, as a Moeid, I believed that behind every universal phenomenon there must be beauty and simplicity in its description. This belief remains true today. On the research side, I finished the requirements for a Master's in Science in about eight months, and was ready to begin research for a Ph.D. degree. All the odds were against my going to America. First, I did not have the connections abroad. Second, the 1967 war had just ended and American stocks in Egypt were at their lowest value, so study missions were only sent to the USSR or Eastern European countries. I had to obtain a scholarship directly from an American University. After corresponding with a dozen universities, the University of Pennsylvania and a few others offered me scholarships, providing the tuition and paying a monthly stipend (some $300). There were still further obstacles against travel to America. It took enormous energy to pass the regulatory and bureaucratic barriers.

Arriving in the States, I had the feeling of being thrown into an ocean. The ocean was full of knowledge, culture, and opportunities, and the choice was clear: I could either learn to swim or sink. The culture was foreign, the language was difficult, but my hopes were high. I did not speak or write English fluently, and I did not know much about western culture in general, or American culture in particular. My presence—as the Egyptian at Penn—was starting to be felt by the professors and students as my scores were high, and I also began a successful course of research. My publication list was increasing, but just as importantly, I was learning new things literally every day—in chemistry, in physics and in other fields. I was working almost "day and night," and doing several projects at the same time. Now, thinking about it, I cannot imagine doing all of this again, but of course then I was "young and innocent". The research for my Ph.D. and the requirements for a degree were essentially completed by 1973, when another war erupted in the Middle East.

I had strong feelings about returning to Egypt to be a University Professor, even though at the beginning of my years in America my memories of the frustrating bureaucracy encountered at the time of my departure were still vivid. With time, things changed, and I recollected all the wonderful years of my childhood and the opportunities Egypt had provided to me. Returning was important to me, but I also knew that Egypt would not be able to provide the scientific atmosphere I had enjoyed in the U.S. A few more years in America would give me and my family two opportunities: first, I could think about another area of research in a different place (while learning to be professorial!). Second, my salary would be higher than that of a graduate student, and we could then buy a big American car that would be so impressive for the new Professor at Alexandria University! I applied for five positions, three in the US, one in Germany and one in Holland, and all of them with world-renowned professors. I received five offers and decided on Berkeley.

Early in 1974 we went to Berkeley, excited by the new opportunities. Culturally, moving from Philadelphia to Berkeley was almost as much of a shock as the transition from Alexandria to Philadelphia—Berkeley was a new world! I saw Telegraph Avenue for the first time, and this was sufficient to indicate the difference. I also met many graduate students whose language and behavior I had never seen before, neither in Alexandria, nor in Philadelphia. The obstacles did not seem as high as they had when I came to the University of Pennsylvania because culturally and scientifically I was better equipped. Berkeley was a great place for science—the BIG science. My general research direction was established, and I immediately saw the importance of the concept of coherence. I decided to tackle the problem, and, in a rather short time, acquired a rigorous theoretical foundation which was new to me. I believe that this transition proved vital in subsequent years of my research. We wrote two papers, one theoretical and the other experimental which were published in Physical Review. These papers were followed by other work, and I extended the concept of coherence to multidimensional systems, publishing my first independently authored paper while at Berkeley. In collaboration with other graduate students, I also published several papers.

During this period, many of the top universities announced new positions, and I was encouraged to apply. I decided to send applications to nearly a dozen places and, at the end, after interviews and enjoyable visits, I was offered an Assistant Professorship at many, including Harvard, Caltech, Chicago, Rice, and Northwestern. My interview at Caltech had gone well, despite the experience of an exhausting two days, visiting each half hour with a different faculty member in chemistry and chemical engineering. The visit was exciting, surprising and memorable. The talks went well and I even received some undeserved praise for style. At one point, I was speaking about what is known as the FVH picture of coherence, where F stands for Feynman, the famous Caltech physicist and Nobel Laureate. I went to the board to write the name and all of a sudden I was stuck on the spelling. Half way through, I turned to the audience and said, "you know how to spell Feynman". A big laugh erupted, and the audience thought I was joking—I wasn't! After accepting the Caltech offer, I was granted tenure in two years and the research group was well established. I never regretted the decision of accepting the Caltech offer.

At Caltech over the years, my science family came from all over the world, and members were of varied backgrounds, cultures, and abilities. The diversity in this "small world" I worked in daily provided the most stimulating environment, with many challenges and much optimism. My research group has had close to 200 graduate students, postdoctoral fellows, and visiting associates. Many of them are now in leading academic, industrial, educational, and governmental positions. Working with such minds in a village of science has been the most rewarding experience—Caltech was the right place for me.

My biological children were all "made in America". I have two daughters whom I am very proud of, Maha, a graduate of Caltech (B.S.) and the University of Texas, Austin (Ph.D.), and Amani, a graduate of Berkeley (B.S.) and currently an M.D. student at the University of Chicago. Dema, my wife, has her M.D. from Damascus University, and completed a Master's degree in Public Health at UCLA. We have two young sons, Nabeel and Hani, and both bring joy and excitement to our life.

The journey from Egypt to America has been full of surprises. As a Moeid, I was unaware of the Nobel Prize in the way I now see its impact in the West. We used to gather around the TV or read in the newspaper about the recognition of famous Egyptian scientists and writers by the President, and these moments gave me and my friends a real thrill—maybe one day we would be in this position ourselves for achievements in science or literature. Some decades later, when President Mubarak bestowed on me the Order of Merit, first class, and the Grand Collar of the Nile ("Kiladate El Niel"), the highest State honor, it brought these emotional boyhood days back to my memory. I never expected that my portrait, next to the pyramids, would be on a postage stamp or that the school I went to as a boy and the road to Rosetta would be named after me. Certainly, I never dreamed that one day I would be honored with the Nobel Prize. But with passion and sincerity, *It Is Possible*, as human achievements should be limited neither by race nor by origin.

SCIENCE IN THE DEVELOPING WORLD

THE FOLLOWING TEXT IS BASED ON A KEYNOTE ADDRESS GIVEN BY NOBEL LAUREATE AHMED H. ZEWAIL (TWAS FELLOW) AT THE TWAS 8TH GENERAL CONFERENCE IN NEW DELHI, INDIA, ON 22 OCTOBER 2002.

I am pleased to have this opportunity to share with you some personal reflections on current issues which I believe may well be at the core of world peace and stability. Science education and development through science are the subject of my presentation and I thought I would use my own journey through two cultures, one currently developing and the other developed, to address issues of concern and what should be done to achieve progress.

As someone who was born and educated in the developing world and who has lived and worked in the developed world, I have acquired both a personal experience and a professional perspective of what it takes to do science at the frontiers of knowledge – not just science for science's sake but science that enlightens the mind and helps our societies and our global community.

In light of the misunderstandings, tensions and violence that plague our world today, I would like to reflect on the critical role that science can play by focusing on issues of concern to the "have nots" of the developing world. After all, they constitute 80 percent of the globe's population.

I was born in Egypt and educated in public schools there. As an undergraduate student at the University of Alexandria, where I earned bachelor's and master's degrees, I was not familiar with such advanced and frontier areas of research as lasers. Later lasers would play a key role in our work that led to the Nobel prize. I did not study advanced quantum mechanics, the language of the microscopic world. Later we would use this language to conduct research on time and matter at the atomic scale. I did not even know much about the Nobel Prize and, as a young boy, I did not spend a single minute dreaming that one day I might receive it.

However, the years that I spent in Egypt as a child and young adult blessed me with a solid foundation for life's journeys based upon these unwavering principles: strength comes from excellence in basic education, especially in science; strong family values are essential for success; and societal appreciation of scholarship motivates youngsters and adults alike.

When I came to the United States, I encountered both cultural and scientific barriers. My English was poor. (In restaurants, I would confusingly ask for "desert" instead of "dessert."). My knowledge of the latest science, especially cutting-edge science, was also poor or at best shaky. There were, moreover, misperceptions of people from my part of the world that often left me distressed. Many thought I rode camels all day long in Egypt. The truth is that I had never ridden a camel in Egypt. Many also imagined that most of my fellow Egyptians drilled for oil all day long. In reality, Egypt has limited oil resources.

Leaving aside these ill-informed views, what the United States gave me was the unique opportunity to develop my potential and a wondrous sense of appreciation for achievement to which I had never been exposed. While learning English and becoming acquainted with the culture, I earned a doctorate degree at the University of Pennsylvania and soon after was appointed a research fellow at the University of California, Berkeley.

My goal, at the time, was quite basic. I wanted to learn and acquire knowledge and then return home, but not before earning enough money to buy a big fancy car – a Buick – to take back to Egypt where I had a permanent teaching position at the University of Alexandria. Through a series of circumstances, however, I eventually wound up as an assistant professor at the California Institute of Technology (Caltech). It was there that I would learn through personal experience the indispensable role that a vibrant culture of science plays in determining the future of young researchers.

There is a widespread misconception in the developing world that progress in science can be driven by buildings and slogans. That is simply not the case. As a youthful untenured professor at Caltech I was given an empty laboratory and some start-up funds. That was all, except for one other thing: enormous freedom to do what I wanted. I did not have a boss. Not even Caltech's president was my boss.

To be sure, after a judicious period of time, my work was assessed but in a thoughtful, yet vigorous, way. Caltech's faculty makes decisions on tenure. After 5 or 6 years, young professors usually know what the final decision is. If things are not going well faculty members shake young professors' hands and wish them good luck in their future endeavours which will take place at a place other than Caltech.

The freedom at Caltech proved special for one simple, yet compelling, reason: It made scholarship – and, more importantly, excellence in scholarship – the driving force on campus.

But the atmosphere was as intimidating as it was exhilarating. The first week after arriving on campus I came in contact with Richard Feynman, Murray Gell-

Mann and Max Delbrück – all Nobel Prize winners. Concluding there was no hope for me, I was ready to pack my bags and return to Egypt. But the scientific atmosphere and culture of science that found its way into every corner of the campus made it clear to me that, if I could develop to the best of my potential, I would likely earn tenure and be on my way to doing all kinds of exciting things in science.

And that is exactly what followed. I received tenure in less than two years. The university appreciated that my research group and I were opening new vistas that created real excitement among scientists around the world. We were not encumbered by bureaucracy. Tens of forms did not have to be signed; tens of seals did not have to be put on paper; and tens of personal status reports did not have to be completed. No push from high-up officials was invoked. A simple, well-defined, transparent system had been put in place – one with sufficient flexibility to ensure that achievement was rewarded fairly, efficiently and effectively.

Why did I earn tenure in less than two years? I headed a research team that came up with something original. Every chemist and physicist working with molecules is interested in their structures. I had a simple idea. I suggested that with lasers and other tools we might be able to look at structural dynamics – that is, structures changing over time and on the time scale of their atomic motions. This is of enormous importance to physical, chemical and biological transformations. If you look at the work of biologists today, they display a like-minded interest as they increasingly focus their research on the relationship between structure and function. Dynamics is at the heart of this junction.

The real challenge was that when scientists do experiments on molecules they look at billions – even trillions – of molecules. How, then, can we examine the intricate

> *I didn't come from Mars with a brilliant idea that would instantly win me a Nobel Prize. It doesn't work that way.*

dynamics of the sheer endless number of molecules (and the even more numerous individual atoms moving inside these molecules)? We turned to lasers trying to develop new techniques with new tools that would bring about the fastest camera for freezing atoms in motion – in a millionth of a billionth of a second.

Much of the conceptual progress, which brought physics and chemistry into confluence, took place at lunchtime in our faculty club, or late at night with my research group members. Each day, I could join informal roundtable discussions. Researchers – young and old, well-known and just starting out – would sit around the table discussing their work. Our research and discussions on many occasions lasted well into dawn. The excitement was everywhere!

The point is that I didn't come from Mars with a brilliant idea that would instantly win me a Nobel Prize. It didn't – and doesn't – work that way. The fanciest building is not responsible for producing breakthrough ideas. What you need is the right scientific atmosphere and the right scientific support.

The work for which I won the Nobel Prize took place in 1987. That was just 10 years after I had arrived at Caltech. The freedom I enjoyed; the camaraderie I experienced; the give-and-take that sharpened my thinking; and the keen awareness I had that my achievements would be recognized and rewarded all helped move my research forward.

That didn't mean our work wasn't scrutinized carefully. Indeed many colleagues initially reacted to our research with scepticism. Critics thought the kind of resolution that we sought (10^{-15} second) would be worthless due to the uncertainty principle that stipulates if you try to do measurements in very brief timeframes you lose information on energy, just as you lose information on the speed of an object, the more precisely you measure its position.

We nevertheless persevered in our work and eventually convinced our colleagues that the uncertainty principle actually operated in our favour. With proof in hand, the criticism was transformed into favourable recognition.

I am convinced that the developing world – even with its limited resources – is capable of producing such an atmosphere. There are scientific centres in the South that have sufficient resources to conduct good research. It's not just a question of money. It's also a question of nurturing a scientific culture that encourages researchers to seek new knowledge and, in the process, challenges them to reach their full potential. Money counts but it must be invested in the right way and not spent on frivolous matters that ultimately have scant impact on the quality of science that is done.

In the past five years, the scientific community worldwide has published about 3.5 million research papers. Europe's share is 37 percent. The US share is 34 percent. The Asia/Pacific share is 22 percent. Other places – representing 70 to 80 percent of the world's population living largely in developing countries – have contributed less than 7 percent of these scientific articles.

What difference does this disparity in academic output make? Should only universities and research centres be concerned? Perhaps not. Consider this interesting correlation. The US contribution to the world's annual economic output is between 30 and 40 percent, comparable to its share of scientific output on a global scale. Europe's annual economic output registers a similar percentage and, like the USA, its economic output tracks its contribution to its output of scientific contributions. It's unlikely that this correlation is coincidental.

If we are aware of these trends and understand the problems that stand in the way of progress, why does the developing world have such difficulties building scientific capacity and putting science to work to improve its economic well-being?

A renaissance in thinking is needed. We need to pay more attention to education and we should invest more in science and technology. We need to lower the political barriers that stand in the way of success and to ensure that our laws do not allow political and fanatical principles to cast shadows over freedom of thought in ways that impede the use of our human resources. Women must participate as full partners in

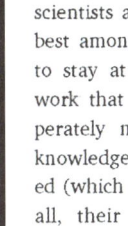

our pursuit of knowledge. The developing world possesses very capable scientists and yet, unwittingly, continues to contribute outstanding scientists to the developed world as part of the brain-drain phenomenon. At Caltech, for instance, my own research group is more than 50 percent Asian.

The developing world, in brief, is rich in human resources. But to take advantage of this invaluable resource, we must develop strategies that nurture and reward the achievements of our scientists and scholars so that the best among them are encouraged to stay at home and pursue the work that their countries so desperately need. While politicians' knowledge of science may be limited (which is understandable; after all, their field is not science), politicians must promote science and its connection to development, and nurture its enormous capacity for interaction on an international scale.

> **The developing world must create new systems of education that emphasize rational thinking.**

The developed and developing world each shoulders responsibilities in efforts to improve the capacity of science in the South and to build better societies that will enable people to enjoy the fruits of science and technology.

First and foremost the developing world must get its house in order. We cannot just wait for the developed world to help us or accuse people there of conspiring against us. Yes, international politics play a role but people's will is a stronger force, provided the force is coherent and not dispersed by internal politics.

Specifically, the developing world must create new systems of education that emphasize rational thinking and that pursues hands-on approaches to the learning of science in ways that engage and excite young students. The objective is to build a new workforce equipped with 21st century tools of education and skills and with a belief in ethics and teamwork. Women must be included in the educational process not only because they deserve to be given an opportunity to succeed but because our societies cannot progress without them. Clearly, this may not be possible on a grand scale in a short time, but the foundation must be established properly and in a timely manner.

Developing countries, moreover, must implement a merit-based system that rewards excellence. Science in much of the developing world relies too much on seniority and puts too much decision-making into too few hands. Everything is centralized and everything needs approval. The result is a snail-paced environment in a fast-paced world.

Such long-standing problems must be addressed in an honest and clear way if we are to go forward. A merit-based system may be the only way to engage and excite young students and to convince them that what they're doing is worthwhile. In the developing world, countries must build their own centres of excellence in science and technology that are especially relevant both to their own country and the global community.

Despite its infrastructural problems, India, for example, has developed centres of excellence that have enabled its scientific community in several disciplines to become partners in international science, education, and a technology-driven economy. But no country can develop centres of excellence unless it creates the right atmosphere for researchers. That means identifying and investing in talent and putting in place a system that minimizes bureaucracy and maximizes freedom and flexibility.

Besides all the obvious benefits of science and technology, the power of knowledge enhances national pride, limits the brain drain and leads the country into effective economical participation in globalization.

The developed world also carries important responsibilities in its efforts to promote scientific capacity and excellence in the developing world. First and foremost it must reform its international aid programmes, investing less money on military hardware and instruction and more on scientific training and partnerships. International aid programmes, moreover, must be drained of politics to ensure money is available for productive North-South initiatives that could help boost science and technology in the developing world.

What will rich countries receive in return for the help they give the "have-nots"?

First, there is the moral dimension. The psychological value derived from being a generous global neighbour should not be underestimated. Even on a personal level, most of us do try to help and all major religions encourage and legitimize efforts to help the needy. It is also difficult to ignore that the prosperity of the North is in part due to natural and human resources from the South and their markets.

Second, we should acknowledge the importance of reciprocation over time. Islamic civilization gave a great deal to Europe, especially during the dark ages. The Arab and Islamic civilizations, which at the time were the world's foremost economic and scientific

powers, were major contributors to the European Renaissance. Today it is the Muslim world that needs help and there is nothing wrong with the United States and Europe (and other developed nations) lending a hand as a modest gesture to the changing fortunes of history.

Third, there is a more practical, self-centred consideration based on the time-tested importance of having an adequate insurance policy. In the United States, I pay a great deal for insurance to protect my family against the high cost of medical care, to protect our house against fire and theft, and to protect our cars against accidents. Similarly, the developed world needs to invest in an insurance policy to help it live in a safer and more secure world.

The choice for the "haves" is clear. They have to help in a genuine and sincere manner. The choice for the "have-nots" is also clear. They first have to get their house in order and build the confidence for a transition to a developed-world status.

In a meeting with Prime Minister Mahathir bin Mohamad on a recent visit to Malaysia, I learned of the critical role that the new education system has played in the nation's rapid transition from a labour-intensive economy dependent on cheap labour to a knowledge-based economy on the doorstep of the developed world. This transition has been fuelled by the belief in building a proper base for science and technology. Malaysia has a majority Muslim population living in harmony with the Chinese and Indian population. Neither religion nor culture seems to hinder progress.

Cheap labour may have worked for developing countries in the past but it will not work in the 21st century. How can the developing world embrace such economy-transforming technologies as microcomputing, genetic engineering, and information technologies without a strong foundation in science? Does the developing world always have to wait decades before participating in global science and technology? Can't we be a part of the modern world without losing our cultural and religious identities? Despite all the political and economic problems we currently face, progress is still possible. But change from within is the first ingredient.

For the sake of global peace and stability the developed and developing world must participate as part-

ners in a dialogue among civilizations and cultures. Such a dialogue should not be confused with slogans theorizing about conflicts between religions or cultures.

At its core, we should nurture a dialogue among the "haves" and "have nots." What's needed is visionary leadership, economic progress, and perspectives that rely on rational thinking. It is for this reason, as much as all the others, that we need global science. ▪

Ahmed Hassan Zewail
TWAS Fellow 1989
Nobel Prize 1999, Chemistry
Linus Pauling Chair
Professor of Chemistry and Physics
California Institute of Technology
Pasadena, California
USA

SCIENCE
and the Search
for Meaning

Perspectives from
International Scientists

જી

Edited by JEAN STAUNE

Foreword by Philip Clayton

Dialogue of Civilizations

Making History through
a New World Vision

AHMED ZEWAIL

The 2002 UNESCO conference, "Science et la quête du sens" in Paris, was devoted to science and the quest for meaning; the English title, "Science and the Spiritual Quest," emphasizes the spiritual dimension, a realm beyond science. Similarly, this chapter,[1] which is based on my lecture given at the conference, is concerned with dimensions beyond science—our human existence in civilizations and cultures that may or may not be in a state of clash. As a scientist, I find these issues complex, but it is precisely this complexity that necessitates a new nondogmatic and rational approach in our quest for human understanding, our search for the truth and new knowledge through science, and our comprehension of the meaning and value of life through faith. My thoughts and reflections are guided by my experience so far in at least three civilizations—the Egyptian, the Muslim-Arab, and the American.

In thinking about the new century and the emerging world, some intellectuals have introduced concepts such as the "clash of civilizations," as termed by Samuel Huntington, and the "end of history," as expressed by Francis Fukuyama.[2] Both authors argue their cases with conviction; nonetheless, these ideas are controversial and debatable. As a scientist, I find no "fundamental physics" to these concepts. In other words, it is not a fundamental principle of civilizations that they be in a state of clash with each other. Neither is it a fundamental principle to end history with one system over all other ideologies.

Here, I argue that the current world disorder results in part from ignorance about civilizations—lack of awareness or selective memory of the past and lack of perspective for the future—and in part from the economic misery and political injustices experienced by the have-nots, which represent some 80 percent of the world's population all across the globe and in *different* civilizations. These are the barriers for achieving the advanced state of world order and, if we can overcome them, we will reach the optimum—a dialogue of civilizations.

Dialogue or Clash?

According to the dictionary, civilization means an *advanced state* of human society in which a high level of culture, science, industry, and government has been reached. Individually, we are civilized when we reach the advanced state of being able to communicate with and respect others of different customs, cultures, and religions. Collectively, we speak of globalization as a means for bringing about prosperity in the world, yet globalization cannot be a practical concept if there are clashes of civilizations. Historically, there are many examples of civilizations that have coexisted without significant clashes.

The central argument in Huntington's thesis is that in this post–Cold War era, the most important distinctions among peoples are not ideological, political, or economic but cultural. He emphasizes the point that people define themselves in terms of ancestry, religion, language, history, values, customs, and institutions; he divides the world into eight major civilizations: Western, Orthodox, Chinese, Japanese, Muslim, Hindu, Latin American, and African.

I have several difficulties with this analysis, and perhaps the following questions and commentary may clarify my position. First, *What is the basis for these divisions of civilizations?* People belong to different cultures, nations have experienced (and continue to experience) different cultures, and nations on the same continent may be influenced by different civilizations. In my case, from birth to the present time, I can identify myself as Egyptian, Arab, Muslim, African, Asian, Middle Eastern, Mediterranean, and American. Looking closely at just one of these civilizations, I note that the Egyptian people belong to a dynamic civilization with a multicultural heritage: Pharaonic, Coptic, Arabic, Islamic, not to mention the Persian, Hellenistic, Roman, and Ottoman influences. The same can be said of the European and American civilizations and others on different continents. The Western cultures of Europe, the United States, and Australia are far from uniform and homogeneous. Given the number of cultures within Europe and the United States, we should then expect a clash of civilizations within a single civilization, without having to look to the other seven. The forces uniting cultures and civilizations are not the result of simple divisions.

A second question is, *Is it fundamental that differences in cultures necessarily produce clashes?* Huntington contends that if the United States loses its European heritage (English language, Christian religion, and Protestant ethics) and its political creed (e.g., liberty, equality), its future will be endangered. From a personal point of view, I did not speak English when I came to the United States; I am not a Christian; and I was not taught Protestant ethics. Yet I integrated myself into my new, American culture while preserving my native culture(s), and I believe that both my "Eastern" and "Western" cultures have benefited from the marriage, without a clash. From a broader perspective, America's strength has traditionally been in its "melting pot"; the country has been enriched—and continues to be enriched—by multi-ethnicity and the different cultures of its inhabitants. As a result, tolerance for different religions and cultures has become part of the American civilization. Provided the people can live in a constitutionally sound system of liberty and equality, intranational clashes are not fundamental; other problems are.

Turning to international relations, it is not obvious to me why civilizations have to acquire their power through imperialism at the expense of the others. Cultures and civilizations can be at their peak of achievement and yet coexist in harmony and even complement each other. The United States, Japan, and European nations are examples of this beneficial coexistence, created by building economic and cultural bridges. The key for achieving this state is to be part of a cooperative world system that represents and observes human liberty and fairness, and whose resolutions are enforced and implemented in a timely manner. This is difficult to achieve, granted, but I believe that visionary leadership can bring it within reach.

A final question is, *What about the dynamics of cultures?* Cultures are not static; they all change with time, and the degree of change is governed largely by forces of politics and economics. Let us consider my home country. Egypt's civilization was developed very early in human history and dominated the world for millennia, but lately the nation has become a developing one. This does not mean that Egypt has lost its civilization, but it does mean that, like others, it has changed with time due to many internal and external forces. In other words, the current state is not due to some intrinsic human or genetic flaw, but rather to the changing fortunes of time.

Other examples of cultural change in Europe and other parts of the world are well known, but the dynamics may be different—different in their time scales and the forces that provoke change. In all cases, however, the dynamics of change cannot be attributed solely to the intrinsic values of an isolated culture. We must take into account political and economic interactions within a culture and between the various cultures of the world. For example, the people of North and South Korea are of similar culture, but the notable disparity in progress between the two countries is due to economic and political factors; the same can be said of East and West Germany before reunification.

The above commentary does not address a problem that is fundamental and common to all cultures and civilizations—the population of have-nots, who have a dynamic of their own. During the European Middle Ages, the peak of Islamic civilization, the majority of Europeans were have-nots, but now most nations of the Muslim world are developing or are underdeveloped, with large populations of have-nots. Some may believe that this is due to a flaw in the intrinsic values of the religion of Islam. It may be useful for me, as an educated Muslim (although I am not a scholar of Islam), to highlight some of the misunderstood principles of Islam and its dynamic civilization. This is also timely given the tragic events of September 11 (2001) in New York and Washington, D.C., their aftermath, and the association in many people's minds of these events with Islam.

Islam and Its Foundations

What is Islam? Islam is the religion and the way of life of about one-fifth of the world's population. There are 1.3 billion Muslims in the world today, 20 percent of whom are Arabs; 5 percent of Arabs are not Muslims. In 1970, there were 500,000 Muslims in the United States; now there are 6 to 7 million, 23 percent of whom are U.S. born. *Islam* is an Arabic word with a double connotation: "peace" and "submission to the will of God." Islam considers itself to be the continuation and the culmination of the earlier "God-sent" religions, Judaism and Christianity; the three are commonly called the monotheistic Abrahamic religions. God commands Muslims to respect all humanity, and Jews and Christians are referred to with distinction as the People of the Book, since they are fellow worshipers of the one God and the recipients of his scriptures (the Torah through Moses and the Gospel through Jesus). The prophet of Islam is Muhammad, who also is the descendant of Abraham through his first son, Ishmael.

Two concepts are basic in Islam[3]: the concept of the unity of God, and the concept of Islam as a way of life, including the civil and legal system. These two concepts are the core of the creed. The Islamic codes of morality are similar to those found in Christianity and Judaism. Muslims accept five primary obligations, commonly called the "five pillars" (*arkan*) of Islam. In practice, of course, Muslims can be seen observing them to varying degrees, for the responsibility of fulfilling the obligations lies on the shoulders of each individual. The pillars are the profession of faith (*shahadah*), prayer (*salah*), almsgiving (*zakah*), fast-ing (*sawm*) during the holy month of Ramadan, and performance of the pilgrimage (*hajj*), the journey to Mecca, for those who can physically and materially afford it, at least once in one's lifetime. Muslims also accept *shariah*, the body of Islamic sacred laws derived from the *sunnah* (custom and religious practice of the Prophet), the *hadith* (documented sayings and teachings of the Prophet), and the Qur'an.

The Qur'an is the scripture of Islam, and Muslims believe it to be authored by God himself and revealed to Muhammad by the Angel Gabriel. The word for God in Arabic is "Allah" and it is used by all Arabs, even Arab Christians and Jews. The Qur'an was revealed in segments of varying length, addressing various issues and circumstances, over the span of twenty-three years, the period of Muhammad's prophethood. Because it is God's direct words, the Qur'an remains in its original language, word for word and letter for letter. Once rendered into any other form or language (even Arabic), it is no longer called the Qur'an, because the direct, divine words are replaced by human words, called interpretations or translations of the meaning. The literary style of the Qur'an is so powerful that to the early Arabs it was an inimitable miracle. The style appears to share features with poetry—again, the Qur'an defies description, being considered neither poetry nor prose but a class unto itself—and this poses difficulty for some non-Muslim readers who like Bible stories told in chronological order. There is one story in the Qur'an (of Joseph) that unfolds chronologically, and to these readers it may still seem poetic.

The Qur'an makes explicit statements about human existence, integrity, and on everything from science and knowledge to birth and death. "Read!" is the first word in the first verse of the direct revelation to the Prophet (Surat al-'Alaq 96:1), and there are numerous verses regarding the importance of knowledge, science, and learning; Muslims position scientists along with the prophets in the respect they are due. The Qur'an provides a general call to humanity: "Cooperate with one another in righteousness and piety, and do not cooperate in sin and transgression" (Surat al-Ma'ida 5:2).

Tragically, some fanatics and some in the media abuse Islam and distort the meaning of its principles through terms such as *jihad* and *terrorism*. The word jihad, for example, is now routinely translated as "holy war," specifically the kind of holy war practiced by Muslims against unbelievers or infidels. This phraseology is far removed from the true concept of jihad in Islam. According to *Lisan al-'Arab*, the most authoritative Arabic dictionary, the word jihad, which derives from the root verb *jahada*, means simply to exert *maximum* effort or striving. The theological connotation of this maximum effort is that it is exerted in striving for betterment—in the struggle within oneself for self-improvement, elevation, purification, and enlightenment. For example, in Egypt, the word *mujtahid* as applied to students means a high achiever. There are other forms of jihad, including the use of economic power to uplift the condition of the needy, and the physical jihad in the struggle against oppression and injustice. The term is also used to denote a war waged in the service of religion. Physical jihad is limited by the following Quranic concepts: "Fight those who fight you, but do not transgress" (2:190); that is, war is justified only if it is defensive in nature. "But if they incline to peace, incline toward it as well, and place your trust in God" (8:61). War is not fought for the purpose of vanquishing or crushing the enemy; peace must be seized at the earliest opportunity. This stress is so important for Muslims that the normal greeting is "Peace be upon you." Islam's peace leaves no room for terrorism, which is the antithesis of jihad. Terrorism is condemned.

A Frustrated Civilization

In general, the West remembers little of the vital role of the Islamic civilization, one of whose centers was in Spain, when Europe was in the so-called Dark Ages. I doubt if the people on the streets of New York, Los Angeles, London, and Paris today are aware of how advanced Islamic civilization was. It provided the world with new knowledge in science, philosophy, literature, law, medicine, and other disciplines. Examples of profound contributions at the turn of the first millennium include those of Ibn Sina, renowned for his work in medicine and known in the West as Avicenna; Ibn Rushd (Averroës) in philosophy and law; Ibn Hayyan (Geber) in chemistry; Ibn al-Haytham (Alhazen) in optics; Omar Khayyam, a renowned poet and mathematician; and al-Khwarizmi, known for his profound contribution to algebra (an Arabic word) and whose name is now commemorated in the word *algorithm*. Bernard Lewis described this civilization well when he traced the history of the region: "For many centuries the world of Islam was in the forefront of human civilization and achievement." He adds, "Islam created a civilization, polyethnic, multiracial, international, and one might even say intercontinental. . . . It was the foremost economic power in the world. . . . It had achieved the highest level so far in human history in the arts and sciences of civilization."[4]

I also doubt that people remember that tolerance was a predominant feature of this so-called Eastern civilization. It was during the peak of the Islamic civilization that Muslims, Jews, and Christians lived together peacefully in Spain and other areas of the Muslim world, and it was in the West that the Jews suffered most from discrimination and torture. Cairo was once the place where Maimonides, the Jewish philosopher, studied the ideas of Avicenna and read Aristotle's work, translated into Arabic by, among others, Christian Arab scholars. Using current events in the world today to ignore the contributions of Islamic civilization and to discredit Islam as intolerant is not conducive to world peace and progress.

Unfortunately, some of the problems facing the Muslim world are the making of Muslims themselves. Many in the Muslim world are not aware of the real message of Islam and some leaders and some fanatics use Islam to enhance their own power and political ambition. Moreover, some create new ideologies in the name of Islam and use their interpretations of the Qur'an in debates to drain the human and intellectual power of the society. I doubt if these people truly understand the meaning of enlightenment and the critical role it played in spreading Islamic civilization, not only among Muslims but also throughout the world at large for nearly a millennium. They also may have forgotten that the Qur'an emphasizes the responsibilities of individuals in improving themselves and their societies, stating, "Indeed! God will not change the good condition of the people as long as they do not change their state of goodness themselves" (al Ra'd 13:11).

Today there is a state of discontentment and frustration in the Muslim and Arab world. These feelings are caused by domestic problems and by global or regional political and economic problems. Because of their glorious past, Muslims are asking, *What went wrong?* As evidenced by past achievements, Islam in its proper state is not a source of backwardness and violence. However, one cannot ignore the influence of modern colonization and occupation by Western powers, the disappointment in the alignment with the Eastern or the Western bloc (communism versus capitalism), which failed to yield prosperity, nor can one overlook domestic problems that often result from the ruling by nondemocratic regimes, in many cases supported by Western governments. Moreover, they see through the world media the dominance and prosperity of the West, the humiliation in Palestine, Bosnia, and Chechnya, and their unfavorable economic status in comparison with the rest of the world.

I do not agree with a conspiracy theory of the West against the East; neither do I believe that all the problems are caused by the West. But I do believe that the West should do more to help, as detailed below. Islamic civilization helped Western civilization in the past and it is reasonable to ask for reciprocation now. Furthermore, new methods for better communication are key to continued progress and coexistence. As discontentment and frustrations grow in the have-not world of more than one billion, the world faces increasing risk of conflict and instability, and such troubles will come from boundaries beyond the Arab and Muslim world.

The World of the Have-Nots

In our world, the distribution of wealth is skewed, creating classes within and among populations and regions of the globe.[5] Only 20 percent of the population enjoys the benefit of life in the "developed world," and the gap between the haves and have-nots continues to increase, threatening our stable and peaceful coexistence. According to the World Bank, out of the 6 billion people on Earth, 4.8 billion are living in developing countries; 3 billion live on less than $2 a day; and 1.2 billion live on less than $1 a day, which defines the absolute poverty standard; 1.5 billion people still do not have access to clean water, with concomitant risk of waterborne diseases; and about 2 billion people are still waiting to benefit from the power of the industrial revolution. The annual per capita gross domestic product (GDP) in some Western, developed countries is $35,000, compared with about $1,000 in many developing countries and significantly less in underdeveloped populations. This factor of up to 100 times the difference in living standards ultimately creates dissatisfaction, violence, and racial and ethnic conflict. Evidence of such dissatisfaction already exists; we have only to look at the borders of developed with developing or underdeveloped countries (for example, in America and Europe) or at the borders between the rich and poor within a nation.

Some believe that a new world order can be achieved through globalization to solve problems such as population explosion, the economic gap, and social disorder. This conclusion is questionable. Globalization, in principle, is a hopeful ideal by which all nations may prosper and advance through participation in the world market. Unfortunately, in its present form, globalization is better tailored to the prospects of the able and the strong, and, although of value to human competition and progress, it serves that fraction of the world's population that is able to exploit the market and the available resources. Moreover, nations have to be ready to enter the gate of globalization, and such entry requires a passage over economic and political barriers.

Barriers to Progress

What is needed to overcome barriers to progress? The answer to this question is not trivial, because many cultural and political considerations are part of the total picture. Nevertheless, I believe that there are essentials for progress that developing and developed countries should seriously consider. For developing countries, there are three essential goals: (1) *building the nation's human resources*, taking into account the necessary elimination of illiteracy, the active participation of women in society, and the need for a reformation of education; (2) *restructuring the national constitution*, which must allow for freedom of thought, minimization of bureaucracy, development of a merit system, and a credible (enforceable) legal code; and (3) *building the science base*.

This last goal is critical for both development and world participation. With a strong scientific base supporting improved education and research, it is possible to enhance the science culture, foster a rational approach, and educate the public about potential developments and benefits. The benefits of science and technology to society are obvious but, just as important, proper science education provides society with rational thinking and thought processes. If absent, a huge void in analytical thinking will be filled with ignorance and even violence. Science is the backbone of progress but, just as important, its knowledge preserves one of the most precious values of humanity—enlightenment.[6,7]

The mindset that such a science base is only for those countries that are already developed is a major obstacle to the have-nots. Moreover, some even believe in a conspiracy theory that the developed world will not help developing countries and that they try to control the flow of knowledge. The former is the "Which came first, the chicken or the egg?" argument, because developed countries were developing before they achieved their current status. Recent success in the world market in developing countries, such as China and India, is the product of their developed educational systems and technological skills in certain sectors—India is fast becoming one of the world leaders in software technology, and products labeled "Made in China" are now all over the globe. As for the conspiracy theory, as stated above, I personally do not give significant weight to it, preferring to believe that nations interact in their mutual interests.

What is needed is acceptance of responsibility in collaboration between developing and developed countries. For the developed world, three essentials are identified:

- *Focusing of aid programs.* Usually, an aid package from developed to developing countries is distributed to many projects (in many cases, most of the aid is for military support). Although some of these projects are badly needed, the number of projects involved and the lack of follow-up (not to mention the presence of corruption) means that the aid does not result in big successes. More direct involvement and focus are needed, especially to help centers of excellence achieve their mission, according to criteria already established in developed countries.
- *Minimization of politics in aid.* The use of an aid program to help specific regimes or groups in the developing world is a big mistake, as history has shown that it is in the best interests of the developed world to help the *people* of the developing countries. Accordingly, an aid program should be visionary in addressing real problems and should provide for long-term investment to ensure true development.
- *Partnership in success.* There are two ways to aid developing countries. Developed nations can either give money intended simply to maintain economic and political stability or they can become partners and provide expertise and a follow-up plan. This serious involvement would be of great help in achieving success in many different sectors. I believe that real success *can* be achieved, provided there exists a sincere desire and a serious commitment to partnerships beneficial to all parties.

Global Returns

What is the return to rich countries for helping poor countries? At the level of the individual, there are religious and philosophical reasons that make the rich give to the poor—morality and self-protection motivate us to help humankind. For countries, mutual aid provides (apart from its altruistic and moral value) insurance for peaceful coexistence and cooperation for preservation of the globe. If we believe that the world is becoming a village because of information technology, then in that village we must provide social security for the less privileged, or we may promote a revolution.

Healthy and sustainable human life requires the participation of all members of the globe. Ozone depletion, for example, is a problem that the developed world cannot handle alone—not only the haves use propellants with chlorofluorocarbons (CFCs). Transmission of diseases, depletion of natural resources, and the greenhouse effect are global issues, and both the haves and the have-nots must address solutions and consequences. Finally, there is the growing world economy. The markets and resources of developing countries are a source of wealth to developed countries, so it is wise to cultivate a harmonious relationship for mutual aid and mutual economic growth.

A powerful example of visionary aid is the Marshall Plan given by the United States to Europe after World War II. Recognizing the mistake made in Europe after World War I, in 1947, the United States decided to help rebuild the damaged infrastructure and to become a partner in Europe's economic (and political) development. Western Europe is stable today and continues to prosper, as does its major trading partner, the United States. The United States spent a mere 2 percent of its GNP on the Marshall Plan from 1948–51. A similar percentage of the $6.6 trillion of the U.S. GNP in 1994 would amount to $130 billion, almost ten times the $15 billion a year currently spent for all nonmilitary foreign aid and more than 280 times the $352 million the United States gave for all overseas population programs in 1991.[8] The commitment and generosity of the Marshall Plan resulted in a spectacular success story. I can see this happening again for Palestine to build a peaceful and prosperous Middle East, and for Africa and Latin America.

It is in the best interests of the developed world to help developing countries sustain a high level of growth to join a new world order and global market. Some developed countries are recognizing the importance of partnerships, especially with neighbors, and attempts are being made to create new ways to support and exchange the know-how; examples include the United States and Mexico and Western and Eastern Europe. The rise of Spain's economic status is in part due to the partnership within Western Europe. By the same token, it is in the best interests of developing countries to address the issues of progress seriously, not through slogans, and with a commitment of both will and resources in order to achieve real progress and to take their places in the developed world.

Building Bridges

Building bridges between cultures and nations is not easy, but the circumstances of the modern world do not permit any culture or nation to remain isolated and insulated. In this century, we are fortunate in having the means to construct such bridges, the mobility to acquire the learning of other cultures, and the human contact that enhances tolerance of other cultures and religions. My own personal experience may be relevant. I am "bicultural." By my fiftieth birthday, I had spent almost equal amounts of time in Egypt and the United States, in the culture of the East and in the culture of the West.

I consider myself fortunate to be enriched by these two cultures, with no culture clash—to gain education in one and contribute to human knowledge in the other, to foster an Eastern tradition in a Western society, and to help facilitate the interaction between the East and the West. This is not new in history. I can envision that the same thing happened when Alexandria, where I received my university education, was a beacon of knowledge. Its famous library, Bibliotheca Alexandrina, brought the West to the East more than two millennia ago.

Science is a universal culture. In the big picture, this universality unites scientists in their search for the truth, no matter what their origin, race, or social background. When I look back at the origins of the science of time and matter, which is central to our research at Caltech, I find a real dialogue. The Eastern, Egyptian civilization I came from was the first to introduce the astronomical calendar around 4240 B.C., measuring accurately the period of a day in a year and, by 1500 B.C., the period of an hour in a day. This was achieved by observing the event of the helical rising of the brilliant star Sothis (or Sirius) and introducing the new technology of sun-clocks or sundials, respectively.

The Western, U.S. civilization I live in gave the world the time resolution of a femtosecond, a millionth of a billionth of a second, the speed needed to record atoms in motion. The concept of the atom, invisible until recently, was given to the world by Democritus of the Greek civilization twenty-five centuries ago. How wonderful and significant that civilizations of different cultures and times have introduced through science enormous benefits to all humanity. It was the rational tradition, in this case of science, that facilitated such building of bridges over millennia of time.

The complexity in world affairs is real and no one can claim that the solutions to world problems are obvious. Whether because of their glorious past or their present geographical and cultural richness, all nations have an important role in helping to solve world problems. As the sole superpower in the world today, the United States has a special role because of its economic, scientific, and military power, but all nations together have responsibilities for a peaceful coexistence in this world.

While the strongest country on Earth must play a fundamental leadership role in combating terrorism together with the international community, it must not lose sight of its leadership role in working for human rights and in reducing the gap between rich and poor, between haves and have-nots. The United States has the opportunity to lead the globe to become a *united* world, to get people all over the world to think of each other as fellow human beings. I vividly remember the American image in the 1960s of a man going to the Moon for the sake of humanity. As Neil Armstrong said in his first words on the Moon: "One small step for man, one giant leap for mankind." The Marshall Plan and the Peace Corps are two examples of visionary initiatives that are representative of that American image of doing great things for humanity.

True, the United States cannot possibly solve every problem in the world, but as the most powerful nation, it should stand tall as a leader and be a role model for others. People around the globe look up to America and many people wish to have an American system of freedom and values. America can be a real partner in helping solve many problems around the world. The reality of the American position was expressed by Zbigniew Brzezinski: "America stands at the center of an interlocking universe, one in which power is exercised through continuous bargaining, dialogue, diffusion, and quest for formal consensus, even though that power originates ultimately from a single source, namely, Washington, D.C."[9]

If history is a coherent and evolutionary process, as argued by Francis Fukuyama, liberal democracy may constitute the end point of humankind's ideological evolution and the final form of human government, and thus it constitutes the "end of history." The argument is supported by the success of the system's economics (free market) and by the successful emergence of the system (democracy) over rival ideologies such as hereditary monarchy, fascism, and communism. This view is controversial, as many believe that Western democracy is not the only viable model of government for the rest of the world; other forms or combinations of systems may be appropriate for different cultures. However, whatever the nature of the system, I believe that human liberty and value, which are basic principles of democracy, are essential for leaps of progress and for the best utilization of human resources. These principles should be exported to the have-nots, but with an understanding of cultural and religious differences, not with hegemony.

Ultimately, with the power of science and technology, and with faith, we will unveil the true nature of our unique consciousness as homo sapiens, the significance of our genetic unity despite race, culture, or religion, and our need for appreciating binding human values. The greatest enemy of human aspiration is ignorance, whether it manifests itself in distorted views of faith, distorted views about other peoples, the failure to recognize the importance and use of new knowledge and new technology, or misunderstandings about nutrition and diseases. It is the source of virtually all human misery.

In this world, we need to build bridges between people, cultures, and nations. Even if we disagree on some issues, these bridges will help us recognize that we live on one globe with common objectives for peaceful coexistence. The key is not to ignore the have-nots, not to ignore the frustrated part of the world. Poverty and hopelessness are sources for terrorism and disruption of world order. Better communications and partnerships will lessen the divide between "us" and "them." We must not allow for the creation of barriers through slogans such as the "clash of civilizations" or the "conflict of religions"—the future is in dialogue, not in conflicts or clashes. We need visionary leaders who make history, not leaders who envision the end of history.

NOTES

1. I wish to acknowledge very useful discussions with Dr. Dema Faham and Dr. Gasser Hathout; Dr. Hathout provided the verses (from the Qur'an) on *jihad*. I also wish to thank Dr. Mary Knight for the careful reading of the manuscript, as well as acknowledge the concise write-up, "Islam: An Introduction," written by the staff of Saudi *Aramco World* (January-February 2002); and, "Glimpses from the Quran," published by the Islamic Center of Southern California (October 2001). Some parts of this chapter are based on presentations I have made in published papers in *Nature* (London), *Proceedings of the Pontifical Academy of Sciences*, and in *Voyage through Time: Walks of Life to the Nobel Prize* (Cairo: The American University in Cairo Press, 2002).

2. Samuel Huntington, *The Clash of Civilizations and the Remaking of World Order* (New York: Simon & Schuster, 1996); idem, "Keynote Address: Colorado College's 125th Anniversary Symposium: Cultures in the 21st Century: Conflicts and Convergences," February 4, 1999; Francis Fukuyama, *The End of History and the Last Man* (New York: Avon Books, 1992).

3. See Karen Armstrong, *A History of God: The 4000-Year Quest of Judaism, Christianity and Islam* (New York: Ballantine Books, 1993).

4. Bernard Lewis, *What Went Wrong? Western Impact and Middle Eastern Response* (New York: Oxford University Press, 2002).

5. Joel E. Cohen, *How Many People Can the Earth Support?* (New York: Norton & Co., 1995).

6. Ahmed Zewail, *The Future of Our World*, in *Einstein — Peace Now! Visions & Ideas*, eds., R. Braun and D. Krieger (Weinehim: Wiley-VCH, 2005, 109); based on the 5th U Thant Distinguished Lecture, United Nations University, Tokyo, 15 April 2003.

7. Ahmed Zewail, *Voyage Through Time: Walks of Life to the Nobel Prize* (Cairo: The American University in Cairo Press, 2002); translated to French, German, Spanish, Chinese, Russian, Korean, and Arabic (appeared in 12 languages and editions).

8. Cohen, *How Many People Can the Earth Support?* Cohen also published "Population Growth and Earth's Human Carrying Capacity," *Science* 269, no. 5222: 341-46; and "How Many People Can Earth Support?" *The Sciences* (Nov./Dec. 1995): 18–23.

9. Zbigniew Brzezinski, *The Grand Chessboard: American Primacy and Its Geostrategic Imperatives* (New York: Basic Books, 1997).

Science for the have-nots

Developed and developing nations can build better partnerships.

Ahmed H. Zewail

Only a fifth of the population enjoys the benefit of life in the 'developed world', and the gap between the haves and have-nots continues to increase, threatening stability. According to the World Bank, of the 6 billion people on Earth, 4.8 billion live in developing countries, 3 billion live on less than US$2 a day, and 1.2 billion live on less than $1 a day, which defines the absolute poverty standard; 1.5 billion people do not have access to clean water.

Although globalization, in principle, aspires to help nations prosper and advance, it is better tailored to the fraction of the world's population able to exploit natural resources and markets. The per capita gross domestic product — the total unduplicated output of economic goods and services produced within a country as measured in monetary terms — has reached $35,000 in some Western countries compared with about $1,000 in many developing countries, and significantly less in underdeveloped populations. For example, according to United Nations statistics: Angola $528; China $777; North Korea $430; South Korea $6,956; United States $31,059; and Yemen $354. This situation will, if left unchecked, exacerbate global instability.

Progress barriers

Developing nations encounter four barriers to achieving developed-world status. High rates of illiteracy in many developing countries reflect the failure of educational systems, and are linked to the alarming increase in unemployment. Second, the limited use of human resources — largely due to hierarchical dominance, strong seniority systems and the centralization of power — suppresses collective human thought and stifles human potential. Third, the mix-up of state laws and religious beliefs causes confusion and chaos through the misuse of religion's fundamental message about the ethical, moral and human ingredients of life. And fourth, there is an incoherent vision for science and technology.

The lack of a solid science and technology base is not always a result of poor capital or human resources. It sometimes stems from a lack of appreciation of the critical role of science and technology in development, an incoherent methodology for establishing such a base, and an absence of a coherent policy addressing national needs, and human and capital resources. Some countries consider scientific progress to be a luxury, in view of other demanding problems. Others believe that the base can be built by buying technology from developed countries. These beliefs

A id programmes should be visionary in their mission and supportive of future investment.

translate into poor or, at most, modest advances that are based on the efforts of individuals, not institutional teamwork.

These issues point to three essential ingredients for progress. First is the building of human resources by eliminating illiteracy, ensuring active participation of women in society, and reforming education. Second is to rethink national constitutions, allowing for freedom of thought, minimizing bureaucracy, developing a merit system, and creating a credible — and enforceable — legal code. Third is the building of a science base.

The foundations of a science base are investment in special education for the gifted, the establishment of centres of excellence, and the chance to apply knowledge in the industrial and economic markets of the country and, eventually, the world. This must go hand-in-hand with a plan for general education at state schools and universities. With such a vision, a scientific culture will emerge that enhances a country's ability to follow and discuss complex problems, rationally and collectively. Scientific thinking becomes essential to the fabric of the society.

Many people feel that a scientific culture is only for developed countries. Some even believe in conspiracy theories — that the developed world will not help developing countries so as to control the flow of knowledge. I do not subscribe to such theories. The recent examples set by China and India, among others, of success in the world market result from their developed educational systems and technological skills in certain sectors. What is needed to develop a scientific culture successfully is acceptance of responsibility in a collaboration between developing and developed countries.

Developing countries need centres of excellence, not only for research and development, but also for training experts in advancing technologies and so reducing the brain drain experienced by many such countries. It is important that these centres are not just exercises in public relations: they should be limited to a few areas in order to build confidence and recognition.

To this end, national resources are needed to support research and development in a selective way, following well-established criteria based on merit and distinction. To guide national policy, government at the highest level should create an overseeing board for science and technology, formed from national and international experts. Without serious commitment to these principles, progress will remain limited.

Basis for success

Some developing countries have made admirable progress in these areas: and the development of India, South Korea and Taiwan reflect this. In Egypt, the University of Science and Technology is involved in an experiment being carried out on a 300-acre site given by the government on the outskirts of Cairo. The university awaits the approval of a new law that will make it a non-profit, non-government organization.

But it is also the responsibility of developed countries to focus aid programmes. Usually an aid package is distributed between many projects, with lack of follow-up leading to diffusion of resources and in some cases corruption, so the aid does not result in significant successes. Real focus can be achieved by establishing what I call 'partnership-guided aid', with a significant fraction of the aid being directed towards excellence using criteria established in developed countries.

There must also be a minimization of politics in aid. The use of an aid programme to help specific regimes or groups is a big mistake, as history has shown that it is in the best interests of the developed world to help the entire populations of developing countries. The aid programme should be visionary in its mission and supportive of investment in future developments. Developed nations either can give money as charity or they can become partners, providing expertise and a follow-up plan.

Some developed countries are recognizing the importance of partnership, especially with their neighbours. Examples include the United States and Mexico, and western and eastern Europe. Real success can be achieved provided there exists a sincere commitment to a partnership. This is in the best interests of both the developed and developing countries for peaceful coexistence in a world of civilized humanity. ∎

Ahmed H. Zewail is in the Divisions of Chemistry and Chemical Engineering; and Physics, Mathematics and Astronomy, California Institute of Technology, Pasadena, California 91125, USA. This commentary is based on his address to the Pontifical Academy of Sciences at the Vatican.

PONTIFICIAE
ACADEMIAE
SCIENTIARVM
SCRIPTA VARIA

99

VATICAN CITY
2001

Science and the Future of Mankind

Science for Man and Man for Science

PROCEEDINGS

WORKING GROUP
12-14 NOVEMBER 1999

JUBILEE PLENARY SESSION
10-13 NOVEMBER 2000

Joannes Paulus PP. II
2000

The Pontifical Academy of Sciences
Casina Pio IV

The Participants of the Plenary Session of 10-13 November 2000

THE NEW WORLD DIS-ORDER – CAN SCIENCE AID THE HAVE-NOTS?

AHMED H. ZEWAIL[1]

On our planet, every human being carries the same genetic material and the same four-letter genetic alphabet. Accordingly, there is no basic genetic superiority that is defined by race, ethnicity, or religion. We do not expect, based on genetics, that a human being of American or French origin should be superior to a human from Africa or Latin America. Moreover, it has been repeatedly proven that men and women from the so-called developing or underdeveloped countries can achieve at the highest level, usually in developed countries, when the appropriate atmosphere for excelling is made possible. Naturally, for any given population, there exists a distribution of abilities, capabilities and creativity.

In our world, the distribution of wealth is skewed, creating classes among populations and regions on the globe. Only 20% of the population enjoys the benefit of life in the "developed world", and the gap between the "haves" and "have-nots" continues to increase, threatening a stable and peaceful coexistence. According to the World Bank, out of the 6 billion people on Earth, 4.8 billion are living in developing countries; 3 billion live on less than $2 a day and 1.2 billion live on less than $1 a day, which defines the absolute poverty standard; 1.5 billion people do not have access to clean water, with health consequences of waterborne diseases, and about 2 billion people are still waiting to benefit from the power of the industrial revolution.

[1] Ahmed Zewail received the 1999 Nobel Prize in Chemistry and currently holds the Linus Pauling Chair at Caltech.

The per capita GDP[2] has reached, in some western, developed countries, $35,000, compared with about $1,000 per year in many developing countries and significantly less in underdeveloped populations. This factor of 40-100 times the difference in living standards will ultimately create dissatisfaction, violence and racial conflict. Evidence of such dissatisfaction already exists and we only have to look at the borders of developed-developing/underdeveloped countries (for example, in America and Europe) or at the borders between the rich and poor within a nation.

Some believe that the "new world order" and "globalization" are the solution to problems such as population explosion,[3] the economic gap and social disorder. This conclusion is questionable. Despite the hoped-for new world order between superpowers, the globe still experiences notable examples of conflict, violence and violations of human rights. The world order is strongly linked to political interest and national self-interest, and in the process many developing countries continue to suffer and their development is threatened. Globalization, in principle, is a hopeful ideal that aspires to help nations prosper and advance through participation in the world market. Unfortunately, globalization is better tailored to the prospects of the able and the strong, and, although of value to human competition and progress, it serves the fraction of the world's population that is able to exploit the market and the available resources.

Moreover, nations have to be ready to enter through the gate of globalization and such entry has requirements. Thomas Friedman, in his book "The Lexus and the Olive Tree", lists the following eight questions in trying to assess the economic power and potential of a country: "How wired is your country? How fast is your country? Is your country harvesting its knowledge? How much does your country weigh? Does your country dare to be open? How good is your country at making friends? Does your country's management 'get it'? and How good is your country's brand?" These

---

[2] Per capita gross domestic product (GDP) in U. S. dollars is the total unduplicated output of economic goods and services produced within a country as measured in monetary terms according to the U. N. System: Angola (528), Canada (19,439), China (777), Hong Kong (24,581), Egypt (1,211), Israel (17,041), North Korea (430), South Korea (6,956), Switzerland (35,910), U. S. A. (31,059), and Yemen (354). From U. N. Statistics Division.

[3] Overpopulation of the world, and its anticipated disasters, is not a new problem. It has been a concern for many millennia, from the time of the Babylonians and Egyptians to this day. Joel Cohen, in his book "How Many People Can the Earth Support?" provides a scholarly overview of the global population problem.

"eight habits of highly effective countries", according to Friedman, are the attributes countries need to succeed in the new era of globalization. The picture in mind is that countries are becoming like global companies, with the aim of being prosperous (in a timely manner). Organization and management and the technical know-how are essentials for a country to prosper; location, history, natural resources or even military might are no longer decisive!

Before attempting to address solutions, it is important to examine the origin of the problem by looking at the "anatomy of the gap". In my view, there are four forces which contribute to the barriers to achieving developed-world status:

Illiteracy: In many countries, especially those in the Southern Hemisphere, the illiteracy rate reaches 40-50% among the general population. Even worse, in some countries, the illiteracy rate among women is above 70%. These rates reflect the failure of educational systems, and are linked to the alarming increase in unemployment. One cannot expect to seriously participate in the world market with this state of unpreparedness. In the west, illiteracy on this scale has been essentially eliminated, and nowadays often means a lack of expertise with computers, not the inability to read and write! Of course, some now developed countries had high illiteracy rates when they began their development, but we must recall that scientific know-how was possessed by a significant portion of the population.

Incoherent Policy for Science & Technology: The lack of a solid science & technology base in the world of have-nots is not always due to poor capital or human resources. Instead, in many cases, it is due to a lack of appreciation for the critical role of science & technology, an incoherent methodology for establishing a science & technology base, and an absence of a coherent policy addressing national needs, human and capital resources (even in some developed countries, we are witnessing the consequences of the latter). Some countries believe that science and technology are only for rich nations. Others consider scientific progress to be a luxury, not a basic need, that it is only necessary to pursue after the country has solved other demanding problems. Some rich, but developing, countries believe that the base for science and technology can be built through purchases of technology from developed countries. These beliefs translate into poor, or at most, modest advances and in almost all cases the success is based on individuals, not institutional teamwork. These complex problems are made worse by the fact that there are many slogans, reports and showcase efforts which do not address the real issues and are intended for local consumption.

Restrictions on Human Thought: Real progress requires the participation of knowledgeable people working together to address key problems and possible solutions. In the west, this participation involves senior and junior people and their different areas of expertise in exchanges of human thought and knowledge. The result is a planned recommendation, designed to help different sectors of the society. In many developing countries, although this practice is true on paper, it is usually not followed in reality. The reasons are many, including hierarchical dominance, strong seniority systems and the centralization of power; all limit people's ability to speak freely. Although western democracies are not the only successful models for government, a lack of democratic participation suppresses collective human thought and limits "due process of the law", which unfairly stifles human potential.

Fanatical Mix-ups of State Laws and Religious Beliefs: Confusion and chaos result from the misuse of the fundamental message of religion, namely the ethical, moral and humanistic ingredients in the life of many, a significant fraction of world population. For example, in Islam the message is clear, fully expressed in the Holy Quran to Muslims, who are close to one billion in global population. The Quran makes fundamental statements about human existence and integrity, on everything from science and knowledge to birth and death. "READ" is the first word in the first verse of the direct Revelation to The Prophet [Sura Alaq 96:1] and there are numerous verses regarding the importance of knowledge, science and learning; Muslims position scientists along with the prophets in the respect they are due. The Quran also emphasizes the critical role that humans must play in the struggle to achieve and develop, stating, *"Verily! Allah will not change the good condition of the people as long as they do not change their state of goodness themselves."* [Sura Al Ra'd 13:11]. All societies and religions experience some fanaticism, but the current disparity in the world economy, with the dominance of the west, and the new role of invading media and politics trigger real fear for the possible loss of religious and cultural values. This situation, with increased unemployment, results in rigidity towards progress and the release of frustration in different ways. The west is seen by many as responsible for some of the mix-up, first because there is inconsistency in political actions by the west, and second because of the gap between the rich and the poor; between billionaires and the homeless.

What is needed to solve these problems? The answer to this question is non-trivial because of the many cultural and political considerations that are part of the total picture. Nevertheless, I believe that the four issues identified above point to the essentials for progress, which are summarized in the following: *(1) Building the human resources*, taking into account the necessary elimination of illiteracy, the active participation of women in society, and the need for a reformation of education: *(2) Rethinking the national constitution*, which must allow for freedom of thought, minimization of bureaucracy, development of a merit system, and a credible (enforceable) legal code; *(3) Building the Science Base*. This last essential of progress is critical to development and to globalization and it is important to examine this point further.

There is a trilogy which represents the heart of any healthy scientific structure: *First, the Science Base*. The backbone of the science base is the investment in the special education among the gifted, the existence of centers of excellence for scientists to blossom, and the opportunity for using the knowledge to impact the industrial and economical markets of the country and hopefully the world. In order to optimize the impact, this plan must go hand-in-hand with that for the general education at state schools and universities. This base must exist, even in a minimal way, to ensure a proper and ethical way of conducting research in a culture of science which demands cooperation as a *team effort* and as a search for the truth. The acquisition of confidence and pride in intellectual successes will lead to a more literate society. *Second, the Development of Technology*. The science base forms the foundation for the development of technologies on both the national and international level. Using the scientific approach, a country will be able to address its needs and channel its resources into success in technologies that are important to, for example, food production, health, management, information, and, hopefully, participation in the world market. *Third, the Science Culture*. Developing countries possess rich cultures of their own in literature, entertainment, sports and history. But, many do not have a "science culture". The science culture enhances a country's ability to follow and discuss complex problems rationally, and based on facts, while involving many voices in an organized, collective manner – scientific thinking becomes essential to the fabric of the society. Because science is not as visible as entertainment, the knowledge of what is new, from modern developments in nutrition to emerging possibilities in the world market, becomes marginalized. With a stronger scientific base, it is possible to enhance the science culture, foster a rational approach, and educate the public about potential developments and benefits.

The above trilogy represents a major obstacle to the have-nots, as many feel that such a structure is only for those countries which are already developed. Some even believe in conspiracy theories – that the developed world will not help developing countries and that they try to control the flow of knowledge. The former is a chicken/egg argument because developed countries were developing before they achieved their current status. The recent examples for success in the world market in countries such as developing China and India and others are because of the developed educational system and technological skills in certain sectors – India is becoming one of the world leaders in software technology and "Made in China" goods are now all over the globe. As for the conspiracy theory, I personally do not give significant weight to it, preferring to believe that nations "interact" in the best of their mutual interests. If the gap is too large, the interest becomes marginalized, but if the gap narrows, the flow of information (including science and technology) becomes easier, even if the two nations involved "do not really have an affinity to each other."

What is needed is acceptance of responsibility in a collaboration between developing and developed countries. I see two sets of responsibilities in what I term a "proposal for partnership". The proposal highlights the following three points for each of them.

Responsibilities of Developing Countries:

(1) Restructuring Education and Science. The force of expatriates in developed countries should be organized and used for help in a serious manner. Expatriates can help the exchange between developed-developing cultures and assist in bringing modern methods of education and research. This will not be successful without the genuine participation of local experts.

(2) Creation of Centers of Excellence. These centers should be limited to a few areas in order to build confidence and recognition and should not be just exercises in public relations. They are important not only for research and development, but also in preparing a new population of experts in advancing technologies. They would also help reduce the brain drain many developing countries experience.

(3) Commitment of National Resources. These resources are needed to support research and development in a selective way, following well-established criteria that are based on merit and distinction. To guide national policy, government at the highest level should create an overseeing "Board for Science & Technology", formed from national and international experts. Without serious commitment to such an effort, progress will remain limited.

Some developing countries have made admirable progress in these areas, and the results from, e.g., India, South Korea, and Taiwan reflect healthy educational reforms and excellence in some science and technology sectors. In Egypt, the University of Science and Technology (UST) is an experiment, initiated with the hope of establishing a center of excellence that will satisfy the criteria of the above trilogy: nurturing the science base, developing technologies important to the region and the world, and fostering the science culture. So far we have had success in structuring an academic foundation and, with the commitment of President M. Hosni Mubarak, we now have a 300-acre parcel available to build a campus on the outskirts of Cairo. We are awaiting the approval of a new law which will position UST as a non-profit, non-governmental organization. This will be a unique experiment where both developing and developed countries can participate, helping a region rich in human capital and potential but in need of peace and prosperity. By the time UST reaches its final stage, it should have satellites benefiting other countries in the area.

Responsibilities of Developed Countries:

(1) Focusing of Aid Programs. Usually an aid package from developed to developing countries is distributed among many projects. Although some of these projects are badly needed, the number of projects involved and the lack of follow-up (not to mention some corruption) means that the aid does not result in big successes. More direct involvement and focus are needed, especially to help Centers of Excellence achieve their mission, and with criteria already established in developed countries.

(2) Minimization of Politics in Aid. The use of an aid program to help specific regimes or groups in the developing world is a big mistake, as history has shown that it is in the best interests of the developed world to help *the people* of developing countries. Accordingly, an aid program should be visionary in addressing real problems and should provide for long-term investment in the development program.

(3) Partnership in Success. There are two ways to aid developing countries. Developed nations can either give money that simply maintains the economic and political stability or they can become a partner and provide expertise and a follow-up plan. This serious involvement would be of great help in achieving success in many different sectors. I believe that real success *can* be achieved provided there exists a sincere desire and serious commitment to a partnership, which is in the best interests of both parties.

What is the return to rich countries for helping poor countries? And what payoff do rich countries get for helping poor countries get richer? These two questions were asked by Joel Cohen in his book mentioned before. At the level of a human individual, there are religious and philosophical reasons which make the rich give to the poor – morality and self-protection motivate us to help humankind. For countries, mutual aid provides (besides the issue of morality): insurance for peaceful coexistence and cooperation for preservation of the globe. If we believe that the world is becoming a village because of information technology, then in a village we must provide social security for the unprivileged, otherwise we may trigger revolution. If the population is not in harmony, grievances will be felt throughout the village and in different ways.

Healthy and sustainable human life requires the participation of all members of the globe. Ozone depletion, for example, is a problem that the developed world cannot handle alone – the use of propellants with chlorofluorocarbons (CFCs) is not only by the haves. Transmission of diseases, global resources, and the Greenhouse Effect are global issues and both the haves and have-nots must address solutions and consequences. Finally, there is the growing world economy. The market (and resources) of developing countries is a source of wealth to developed countries and it is wise to cultivate a harmonious relationship for mutual aid and mutual economic growth. I heard of a recent phrase, "Give us the technology and we will give you the market!", used to describe the US-China relationship.

A powerful example of visionary aid is the Marshall Plan given by the United States to Europe after World War II. Recognizing the mistake made in Europe after W.W.I, the U. S. decided in 1947 to help rebuild the damaged infrastructure and to become a partner in the economical (and political) developments. Western Europe is stable today and continues to prosper – likewise its major trading partner, the United States of America. The U. S. spent close to 2% of its GNP on the Marshall Plan for the years 1948-51. As pointed out by Cohen, a similar percentage of the $6.6 trillion of the 1994 U. S. GNP will amount to $130 billion, almost ten times the $15 billion a year currently spent for all non-military foreign aid and more than 280 times the $352 million the U. S. gave for all overseas population programs in 1991. The commitment and generosity of the Marshall Plan resulted in a spectacular success story. The world needs a rational commitment to aid and aid partnerships.

It is in the best interest of the developed world to help developing countries become self-sufficient and a part of the new world order and market.

Some developed countries are recognizing the importance of partnership, especially with neighbors, and attempts are made to create new ways of support and exchanges for the know-how. Examples include the United States and Mexico and Western and Eastern Europe. The rise of Spain's economic status is in part due to the partnership within Western Europe.

In the next 25 years, 2 billion human beings will be added to the planet, with 97% of those 2 billion people living in the developing world. This uneven population explosion, with its impact on world resources, the environment and regional conflicts, threatens our existence and calls for serious and active involvement. The consequence when developing countries acquire an "underdeveloped status" is ugly, not only because of the human costs and sufferings, but also because of the impact on world peace and stability. It is equally in the best interests of developing countries to address these issues seriously, not through slogans, but with a commitment of both will and resources in order to achieve real progress and to take a place on the map of the developed world.

We may picture the current situation by likening it to a "Ship in a Flood". Underdeveloped countries are near to sinking under the deluge; developing countries are trying to make it onto the ship; and developed countries sailing, *but* in a flood of the unprivileged. The choices are clear: The sailing ship must seriously attempt to help those who are trying to make it. Those trying to make it should not regard the ship without a willingness to put forth their own effort, and without wasting their energy on conspiracy theories – being on the ship is more important! Meanwhile, everyone must make every attempt to rescue those at the bottom. To be part of a civilized planet, every human must matter. The notion of "us" and "them" is not visionary and we must speak of global problems and solutions. At the heart are poverty, illiteracy, and human freedom.

Chapter 4

The Soft Power: Science & Technology

Dreaming the Future[*]

commentary

Dire need for a Middle Eastern science spring

Ahmed H. Zewail

The Middle East is rich in human and natural resources, but many of its countries need a cultural and scientific transformation to reach worldwide recognition in education, research and economic productivity. Several institutions are making a positive impact, kindling hope for a successful 'science spring'.

The outcome of the uprisings across the Middle East and North Africa may be uncertain, but what is clear is that a political transformation has taken place. People in countries such as Egypt and Tunisia will not allow a return to totalitarian governance. Not only can they demonstrate and bring down governments — they will no longer tolerate a degraded economic and educational status. Through social media and the Internet, they 'see' the world and ponder why they have not achieved what their counterparts in South Korea or China have. Pundits may argue that the outcome of these uprisings should be democracy, but equally important are the scientific and cultural transformations that are essential for development and diplomacy to flourish. Let us recall that in the evolution of Western civilization, the Enlightenment came ahead of modern democracy, and both before the current governance structure and social order.

I have been concerned with these issues in the Middle East for decades — in official and non-official capacities — and their relevance to a 'science for the have-nots' vision, which addresses the mission of aid programmes

and diffusion of science in developing countries[1]. In November 2009 I was asked to be the first US science envoy to the Middle East and soon after I began the inaugural mission, visiting Egypt (the most populous country in the Arab world at 85 million with a gross domestic product (GDP) of US$6,500 per capita), Turkey (80 million people of Middle Eastern, but non-Arab, descent and a GDP of US$15,000 per capita) and the Gulf state of Qatar (2 million people with nearly 0.3 million Qataris and a GDP of US$100,000 per capita). Figure 1a,b shows the total population, literate population and GDP of these countries, as well as those of Iran (whose population is similar to Egypt) and South Korea (an Eastern Asian country with a GDP remarkably higher than that of Egypt, and which has shown a significant scientific development in the past few decades).

The meetings were broad-ranging and included visits with government officials (heads of state, prime ministers, ministers and some members of parliament), members of the education sector (teachers, students and university professors), institutions of higher education and research (private and

state universities), members of the private sector (economists, industrialists, writers and publishers) and some media representatives. These visits exposed the plight of education and science in the region, which lags behind international standards (the consequences are clearly spelled out in the 2003 Arab Human Development Report sponsored by the United Nations Development Programme[2]). The data are telling: whereas the expenditure in research and development (R&D) of South Korea and Israel reaches 4% of the GDP and both countries spend 5% of their similar GDP (US$30,000 per capita) in education, Egypt's R&D expenditure is 0.4% at a GDP of US$6,500 per capita (Fig. 1b,c). The current situation is in sharp contrast to the system of schools and universities that existed even in the 1960s, when I benefitted personally from an excellent education in Egypt.

In the Middle East, Israel leads today in scientific impact, and a major part of its GDP is science-driven. In general, publication and citation indicators show some encouraging trends for the region over the past decade (Fig. 2). However, the impact of scientific research in the Arabian, Persian and Turkish

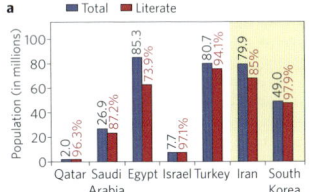
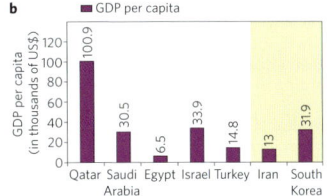
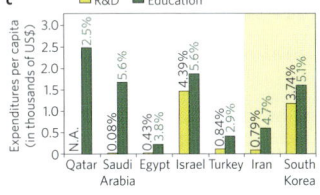

Figure 1 | Demographic and economic statistics for some Middle Eastern countries and, for comparison, for Iran and South Korea. **a**, Total and literate population. Percentages of literate people with respect to total population numbers are also shown. **b**, GDP per capita. **c**, Expenditure per capita in R&D and education. R&D (black labels) and education (green labels) expenditure in percentage GDP are indicated. Population (ref. 16) refers to 2013. Literacy rate (ref. 16): Qatar (2010), Saudi Arabia (2011), Egypt (2012), Israel (2004), Turkey (2011), Iran (2008) and South Korea (2002). GDP per capita (ref. 16) refers to 2012. Percentage GDP expenditure on R&D (ref. 17): Qatar (not available, N.A.), Saudi Arabia (2009), Egypt (2011), Israel (2011), Turkey (2010), Iran (2008) and South Korea (2010). Percentage GDP expenditure on education (ref. 16): Qatar (2008), Saudi Arabia (2008), Egypt (2008), Israel (2009), Turkey (2006), Iran (2010) and South Korea (2009). Raw data for a comparison among countries for exactly the same time periods are not available.

Middle East still pales to that of Israel and the West, and it is natural to ask why the scientists have, as a group, underperformed compared with their colleagues in the West, or with those rising in the East. The reasons are numerous. In the Arab world they include the illiteracy pile-up during colonization, the poor governance and imposed deprivation of free thinking over the past half century, and the continued deterioration of core education owing to the ineffective policies of handling a large student population.

Beyond the causes of the current state of affairs, it is important to understand what is being done to change this situation. The approach varies, depending on country. Here, I shall consider four centres, highlighted together with their geographical positions in Fig. 3, which are already operating in countries where Islam as a culture is the major religion.

In the Gulf, I have served on the Board of Directors of the Qatar Foundation for nearly a decade, during which time I witnessed the birth of a new experiment in university learning; namely, the transfer from the West of established university systems offering cutting-edge education curricula to students of the region. Today, there exist many reputable schools that are sponsored by the Qatar Foundation, and these include Carnegie Mellon University for degrees in computer or biological sciences, Weill Cornell College for medical degrees and Georgetown University for degrees in international affairs or international economics[3].

These and other universities are granting degrees in Doha marked with the institutional insignias of their Pittsburgh, New York and Washington DC campuses. The experiment is very costly to Qatar, but for its purpose it is a successful one, as it brings a new culture of learning to this and nearby countries and defines new standards in higher education. At the moment these institutions serve a relatively small population, but with time the opportunity exists for a much larger student body, as both space and funds have already been endowed for the foundation.

At the graduate level, two institutions in Turkey and Saudi Arabia provide different structures and may make scientific impact by different means. Bilkent, or City of Science, is the first private university in Turkey[4]. As a 30-year-old institution, it has established itself as a leader in undergraduate education and graduate research, and is ranked as one of the best educational institutions in Eurasia. While visiting Bilkent, I was impressed by the high density of world-class faculty, the culture of the place and the drive to achieve at the highest scientific level. Turkey has the necessary human resources and Bilkent is able to attract the best students and faculty.

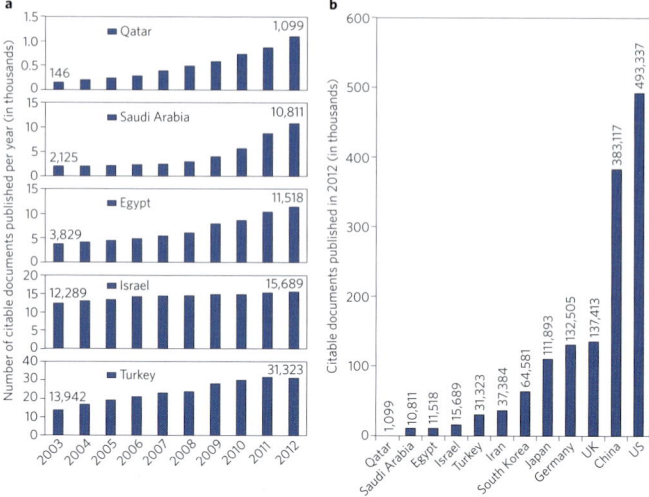

Figure 2 | Scientific production in the Middle East. **a**, Ten-year trend of the number of citable documents published in some Middle Eastern countries. **b**, Publication output in 2012 of the countries indicated in **a** as well as Iran, South Korea and the world's top five countries as per publication output. Data source: ref. 18.

King Abdullah University of Science and Technology (KAUST) in Saudi Arabia is another institution in the region that focuses on research and development[5]. The vice president of KAUST, Jean Fréchet, provides in *Nature Materials* the perspective of the administration and what is hoped for in the future[6]. More than US$10 billion has been committed to this Saudi project. Most of the staff and administration are brought in from outside the Kingdom. Surely there will be research output in the coming years, in diverse and important areas, with new opportunities. The real challenge is how to disseminate this new science culture throughout the country, beyond the premises of KAUST, and how to make it attractive enough to Saudi nationals so as to secure continuity and sustainability.

In Egypt, a different course has been taken — one that I have personally charted for more than a decade. The concept of the project was outlined in 1999 to the then president of the republic, but owing to political reasons, implementation was halted. After the revolution of January 2011, the project was revived and the Egyptian government decreed its establishment as the national project for scientific renaissance, naming it Zewail City of Science and Technology[7]. The City was inaugurated in November 2011 on a campus on the outskirts of Cairo.

The project is unique in several respects. First, in an unprecedented way — certainly in Egypt — the City is supported by donations

from the Egyptian people and from the government. Second, the City has its own special law, granted in 2012, which allows for independent governance by a board of trustees that includes five Nobel laureates. Third, the City comprises three interactive substructures: the university, the research institutes and the technology pyramid, designed to enable world-class education, scientific research and industrial impact. The goal is to build a modern science base with an advanced industry sector and, as importantly, to limit the brain drain in these advanced fields of science and engineering.

The prime purpose of the university is to attract talented students from all over the country and to offer them unique academic curricula that are tailored to provide knowledge in cutting-edge fields of science and engineering. The new concept here is the departure from traditional departments with walls separating the disciplines. Rather, all students have the opportunity to learn in a multidisciplinary, or transdisciplinary, system. This year, 6,000 students applied for admission and 300 were admitted — a 5% admission rate that is on a par with Harvard and Yale.

The second branch of the City, the research institutes, houses centres in fields at the forefront of science and engineering. Priority is given to research particularly pertinent to national needs. The scope of research is broad, from biomedical sciences — which

commentary

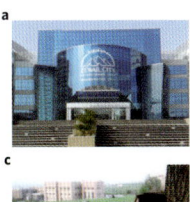

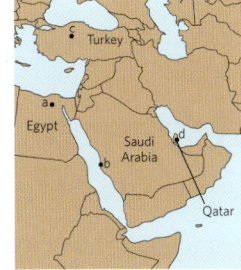

Figure 3 | Research and education institutions in the Middle East. **a**, Zewail City. **b**, KAUST. **c**, Bilkent. **d**, Qatar Foundation. **e**, Geographical position of these institutions. Images from: **a**, © Zewail City of Science and Technology; **b**, © Flickr Vision/Getty; **c**, © Caro/Alamy; **d**, © Philipus/Alamy.

are important for alleviating diseases of the region — to R&D in areas such as solar energy (an alternative source abundant in Egypt). At present, we have seven centres of research in fundamental physics, materials science, nanotechnology, imaging and biomedical sciences, among others. The last substructure is the technology pyramid, whose purpose is to transfer the output of the research institutes to industrial applications, to initiate incubators and spin-off companies, and to attract major international corporations.

These centres of excellence in the region have the potential to transform the overall state of science and the culture of learning. However, in countries rich with human capital but poor in governance and literacy, other revolutionary changes must be implemented[8].

First, the eradication of illiteracy and the building of human capital with the participation of women in the work force are paramount. The current education system is based on rote learning with a focus on quantity rather than quality and this must be replaced with a system that is merit-based and aimed at encouraging free and creative thinking.

Second, the reform of the constitution is essential to allow for freedom of thought and the insulation of scientific enquiry from political and religious interference. The constitution should also state that governments must increase funding of R&D to at least 1% of the GDP and decrease bureaucracy by reforming obstructive laws and regulations.

Third, and perhaps most difficult, is cultural reform. After decades of poor governance and religious conflicts, there is a dire need for a change in thinking, from intolerance to tolerance of others and their opinions, or simply towards understanding the virtues of pluralism[9]. Without a healthy education system and enlightened centres of

excellence, the hope of making such changes is dimmed by the use of political manoeuvres and religious hindrances.

Egypt was and still is the leader of the Arab world. Its next revolution in education and culture will trigger major changes in other Arab countries[10]. Even though this leadership role has slipped over the past three decades during Mubarak's reign, Egypt still has the history and the foundation, not to mention the population and institutions, to regain the avant-garde force of the necessary transformation. Egypt pioneered democratic governance in the region through parliamentary elections, a practice the country is struggling to continue today. More than a century ago, numerous industries emerged including banking, mass media — such as the well-known *Al-Ahram* newspaper — textiles, motion pictures and others. With such achievements, Egypt at that time was ahead of South Korea and Japan in management, education and related fields. Today Egypt is the home of Al-Azhar University (which is older than Oxford and Cambridge) and Cairo University (a centre of enlightenment not only in Egypt but also for the whole Arab world). So the potential for change and rebuilding is real.

The Middle East, with all its resources, is no less capable of development than Southeast Asian countries. The human capital is available and a major fraction of the population is young (nearly 70% of the population is under 30 years of age[9,11]). The region is also rich in natural resources, enjoys a moderate climate and the people are culturally inclined towards hospitality and commerce.

The deficits of the region — which the 2003 Arab Human Development Report identifies as freedom, knowledge and gender gap[2] — can be remedied through education and governance reforms, as outlined above. Education and research reforms must go

beyond the failed patching policies that have been made in past years[12]; they must foster creative thinking and innovation as well as encourage curiosity-driven research[13,14]. Furthermore, they must instil the meaning of citizenry, which will not be acquired without appreciation for the value of free debates, team work and respect for pluralism.

Arab awakening is a reality, but political conflicts in the region obscure its potential. Here, also, education and science can play a key role in diplomacy and the acceleration of the peace process. Once Arabs realize scientific achievements through a transformative educational and cultural process, and when countries such as Israel realize the potential of such achievements, a more rational and serious dialogue for diplomacy and a comprehensive and just peace will divert the energy of the region towards human and economic developments.

The hope is that the political awakening already in motion in the region will support a successful 'science spring' that sweeps the Middle East and enables the building of a knowledge society. Knowledge acquisition is a concept that is woven into the fabric of Islam and was the springboard of success of its empire centuries ago[15]. But to regain prominence in today's world, that concept must re-emerge, transforming the culture in ways necessary for the charting of a new and promising future.   ❐

Ahmed Zewail is at the California Institute of Technology, Pasadena, California 91125, USA. He is also the Chairman of the Board of Trustees of Zewail City of Science and Technology in Egypt. e-mail: zewail@caltech.edu

References
1. Zewail, A. H *Nature* **410**, 741 (2001).
2. United Nations Development Programme *Arab Human Development Report 2003: Building a Knowledge Society* (United Nations Publications, 2003); http://www.arab-hdr.org/publications/other/ahdr/ahdr2003e.pdf
3. http://www.qf.org.qa/home
4. http://www.bilkent.edu.tr
5. http://www.kaust.edu.sa
6. Fréchet, J. M. J. *Nature Mater.* **13**, 321–322 (2014).
7. http://www.zewailcity.edu.eg
8. Zewail, A. H. & Sedky, S. *Nature Middle East* http://dx.doi.org/10.1038/nmiddleeast.2014.5 (2014).
9. Muasher, M. *The Second Arab Awakening and the Battle for Pluralism* (Yale Univ. Press, 2014).
10. Zewail, A. H. The revolution Egypt needs. *The New York Times* (13 October 2013); available via http://go.nature.com/HE379y
11. World Bank *Recovering from the Crisis. World Bank Middle East and North Africa Region: A Regional Economic Update* (The World Bank, April 2010).
12. Laggards trying to catch up. *The Economist* (15 October 2009); http://www.economist.com/node/14660446
13. Zewail, A. H. *Nature* **468**, 347 (2010).
14. Zewail, A. H. How curiosity begat Curiosity. *Los Angeles Times* (19 August 2012); available via http://go.nature.com/lqVzDm
15. The road to renewal: After centuries of stagnation science is making a comeback in the Islamic world. *The Economist* (26 January 2013); http://www.economist.com/node/21570677/print
16. Central Intelligence Agency *The World Factbook*; https://www.cia.gov/library/publications/the-world-factbook
17. http://data.worldbank.org
18. http://www.scimagojr.com/countryrank.php

nature **MIDDLE EAST**
Emerging science in the Arab world

Published: 9 January 2014

Science in Arab Renaissance

While recent events in the Arab world have focused on political upheaval, the region is now in dire need of a new revolution to reform the cultures of education and research.

Commentary by Ahmed H. Zewail and Sherif Sedky

Ahmed Zewail

In the past few years, an awakening, through the 'Arab Spring', has focused on a political dimension of societal change. While the process of transformation begins with democracy, it does not end there. Though public uprisings have brought political changes, a new revolution is needed to transform the culture of learning. The failure of Arab education is a significant underlying cause of youth discontent in the region and has serious cultural, economic and political consequences.

The status of the Arab world in science and education is unacceptable. Its contribution to international scientific research is insignificant and Arab universities do not regularly rank among the world's 500 best institutions. It is remarkable that between 25 and 40% of the Arab population of 350 million remains illiterate, while adult skills in the digital age are now defined in terms of literacy, numeracy, and problem solving.

In Egypt, which has the largest population of Arab countries, hundreds of thousands of students get a university education that is not compatible with the modern world. And, on the global market, there are no technological products "made in Arabia".

It is too simplistic to ascribe a single cause, such as a false

Sherif Sedky

distinction between faith and reason. From a genetic point of view, Arabs are no different from those of any other ethnicity; there is no geographic monopoly on intelligence. Clearly, Arabs and Muslims in Spain, North Africa, and Arabia were at the apex of civilization when Christian Europe was in the dark ages.

The reasons for this lack of endeavor and achievement are myriad; including colonization, corruption, and constitutional deficiencies restricting human liberty and freedom of thought. And, for decades, the use of religion in politics and politics in religion have clouded national goals and diverted attention from the real issues facing Arab nations.

Renaissance in the Arab world will not be possible without genuine government recognition of the critical role of science in development.

The question that needs to be answered is not what went wrong, but what can be done now? Revolutionary changes, not incremental ones, have to take place in education and scientific thought, with three essential components for progress.

First is the building of human resources by restoring literacy, ensuring active participation of women in society, and reforming education.

Second, there is a need to reform the national constitution to allow freedom of thought; to streamline and rationalize bureaucracy; to develop merit-based systems and create a credible and enforceable legal code.

Thirdly, and most tangibly, the constitution should ensure that the budget for research and development is above 1% of the nation's GDP. In light of recent revolutions in Egypt, Tunisia and elsewhere, these changes are possible.

In this vision, human capital is paramount. The Arab world needs to nurture a new generation of professionals capable of thinking critically and creatively, one with contemporary knowledge of science and technology, and with the new emerging disciplines in physical, medical, and social sciences.

Such a pool of knowledge would help identify and provide solutions for fundamental problems facing society. Research into alternative energies, water resources, or drug design can bring numerous social benefits and rewards from the country's economic growth and the participation in the global market.

Changes must be initiated at the root of the education system which is based on rote learning with a focus on the quantity rather than the quality of information delivered to students. This should be replaced with a merit-based system designed to encourage free and creative thinking, with practical experiences.

The landscape of funding and recognition in research also needs to be reformed. Today's convention of using the number of publications to determine candidates for academic promotion has proven to be of little value and must be replaced with a framework to identify original and innovative contributions. Finally, there should be a strong link forged between the academic and industrial sectors to maximize potential mutual benefits from basic research and industrial interests, globally and domestically.

In Egypt, the pioneering National Project for Scientific Renaissance, "Zewail City of Science and Technology", was set up in 2011 to encompass these concepts in education, research, and the industrial impact. The project is funded by donations from the Egyptian people and the government.

At the heart of the City is the University of Science and Technology, whose primary role is to attract talented students from all over the country and offer unique academic curricula in cutting-edge fields of sciences and engineering.

The second major arm of the project comprises research institutes staffed with world-class scientists prioritizing research into essential national problems. The institutes, equipped with state-of-the-art equipment, represent a wide range of interdisciplinary fields, including nanotechnology, environmental engineering, renewable energy, space and communication technology, materials science, biomedical science, and physics of earth and the universe.

The third and last major component is the "Technology Pyramid", which is responsible for commuting research output to industrial applications. It is designed to establish, with intellectual property protection, incubators and spin-off companies and to attract major international corporations to encourage a healthy climate for research-industry exchanges.

The benefits from the Zewail City of Science and Technology, domestically and on a world stage, are numerous. We believe this prototype initiative of three structures, if transferred to other parts of the Arab world, will change the landscape of education and scientific research, and, through significant international participation, will foster new opportunities for the youth of Arab nations.

Renaissance in the Arab world will not be possible without genuine government recognition of the critical role of science in development and policies providing commensurate funding for basic research and reform of rigid bureaucracy which thwarts progress.

In doing so, Arab nations will regain confidence to compete in today's international science and globalised economy. It is gratifying to see several new centres of advanced education and research and development being set up in the region.

However, the goal of the Egyptian initiative is different. Under one umbrella we are providing the milieu for education and research, from school and university stages to advanced manufacturing and market levels. This may be a route to developing society scientifically and culturally, to growing economically, and for restoring the prominent role of Arabs worldwide.

Ahmed Zewail won the 1999 Nobel Prize in Chemistry. He is the Linus Pauling Chair professor of chemistry and professor of physics at the California Institute of Technology (Caltech), and is currently the director of the Moore Foundation's Center for Physical

Biology at Caltech. He is also the founding president of the Zewail City of Science and Technology. Sherif Sedky is the founding provost of Zewail University of Science and Technology and the director of the Nanotechnology Center at Zewail City of Science and Technology.

Correction: Dr. Zewail is not the founding president of Zewail City, but rather the Founder and first Chairman of the Board of Trustees.

MARCH 28, 2011

C&EN

CHEMICAL & ENGINEERING NEWS

HONEYBEE HEALTH
Treatments help colonies
fight pests, viruses **P.24**

PESTICIDE RULES
Farm-state lawmakers
scrutinize EPA actions **P.32**

2011 PRIESTLEY MEDAL
ACS honors Ahmed H. Zewail P.12

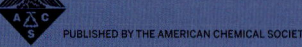

PUBLISHED BY THE AMERICAN CHEMICAL SOCIETY

SCIENTIST-DIPLOMAT EXTRAORDINAIRE

Straddling the U.S. and Middle East, **PRIESTLEY MEDALIST AHMED H. ZEWAIL**
thrives at the interface of science, culture, and international affairs

MITCH JACOBY, C&EN CHICAGO

EVEN BY AHMED H. ZEWAIL'S standards, the intensity of the past month or so has been off the charts.

Zewail, the Linus Pauling Professor of Chemistry and professor of physics at California Institute of Technology, is no stranger to keeping a relentless and high-profile work and meeting schedule. What with the demands of leading a busy research group, traveling the globe to lecture about science and education, and meeting over the years with presidents, prime ministers, kings, queens, and even the pope, Zewail—who won the 1999 Nobel Prize in Chemistry and now serves in President Barack Obama's Administration as the Middle East science envoy—has seen plenty of headline-making action.

Nonetheless, this year's Priestley Medalist, who is known internationally for his pioneering femtosecond spectroscopy work and for ultrafast electron microscopy imaging, says the history he personally participated in and witnessed unfolding before him last month in his native Egypt "is by any measure, the most momentous event in my life."

Even before millions of demonstrators took to the streets of Cairo, Alexandria, and other Egyptian cities in late January and early February, calling for longtime Egyptian president Hosni Mubarak to step aside, Zewail had been planning to spend time in Egypt. As Middle East science envoy and a member of President Obama's Council of Advisors on Science & Technology, Zcwail had rcprescnted the U.S. in the past year on trips to Egypt, Qatar, Dubai, and Turkey. His schedule called for him to return again to Egypt to continue promoting scientific and educational collaborations designed to foster friendlier relations

COVER STORY

DESTINED TO SUCCEED Even as a youngster, seen here dressed for a formal photo at age 12 (left) and at age 10 relaxing on the beach in Alexandria with his father (below), Zewail was motivated to one day become "Dr. Ahmed," as the sign on the bedroom door of his youth predicted.

COURTESY OF AHMED ZEWAIL (BOTH)

in support of U.S. diplomacy in that region.

But when the popular uprising began to take hold in Egypt, Zewail sprang into action—not as a U.S. government representative, but rather as a concerned citizen.

"I made it very clear that I was coming to Egypt as an Egyptian citizen to help my mother country," says Zewail, who holds dual citizenship.

Before leaving for Egypt, he quickly prepared statements that were broadcast on Al Jazeera, Al Arabiya, and Egyptian television, and he wrote opinion pieces that were published in the *New York Times* and the *Times of London* declaring that superficial and cosmetic changes to Egypt's power structure simply wouldn't cut it. Fundamental change is what's needed, Zewail said via those media outlets, and he respectfully but directly urged Mr. Mubarak to step down immediately.

UPON ARRIVING IN Cairo, Zewail kept a fast-paced schedule as he continued to deliver that message in public forums. "I was mainly trying to give moral support, especially to Egypt's energized youth, who simply thirst for democracy and a new Egypt," he says. He did so by holding interviews with numerous news organizations, including BBC and CNN, and appearing on CBS News's "Face the Nation" and "PBS NewsHour." In addition, he served as an unofficial mediator, spending long hours meeting with various youth leaders, protest organizers, activists, and young academics and presenting their views to high-ranking ministers and to Omar Suleiman, who briefly held the post of Egypt's vice president last month.

As the events of early February played out in Cairo, and especially as news of Mubarak's resignation on Feb. 11 sparked major celebrations, the media was abuzz with speculation about who might become Egypt's next president.

Journalists, bloggers, and even science enthusiasts who were closely following the story, drew up short lists of likely candidates that often included

Zewail's name.

After a meeting that Zewail held with young Egyptian researchers, *Nature Middle East* posted a blog entry announcing that Zewail had reconsidered his oft-stated intention to remain in science and stay away from high office—and was now thinking about running for the presidency.

But in a recent interview with C&EN, Zewail emphasized that he does not really want to become president of Egypt. "I have always said I can serve both Egypt and the world more effectively as a scientist, and that is my preference," he asserted. Yet out of a sense of duty and respect for his fellow Egyptians, Zewail is cautious not to reject too hastily the barrage of requests he says he has received by phone, e-mail, and in person, urging him to consider becoming Egypt's president.

As the Arab world's only science Nobel Laureate, Zewail is accorded rock-star status in the Middle East. Students and young people pack auditoriums to hear him speak. For example, last July, more than 5,000 people gathered at the Alexandria Library to hear Zewail speak, according to newspaper and blog accounts of the event. One of his televised addresses, delivered last year from Cairo and broadcast across the Arab world, drew some 30 million viewers, Zewail says, citing information given to him by the satellite station's chief executive officer. Margaret Warner of "PBS NewsHour" recently likened interviewing Zewail in downtown Cairo to "sitting in public with Bono and Einstein

PRIESTLEY
MEDAL
2011

AHMED H.
ZEWAIL

combined," referring to the lead singer of Irish superband U2.

The world of Islam holds intellectual achievement in very high regard, Zewail says, explaining the likely source of his popularity. Underscoring that point, he quotes a lofty Arabic expression, *al-'ilm nur*, meaning "knowledge (or science) is light," and another one comparing *'ilm*, knowledge or science, to water and air, the essentials of life. So in a culture with a tradition of praising academic excellence so highly, especially one that in recent memory has witnessed so few of its members rise to international prominence for scholarship, Zewail stands out.

BUT WITH that popularity comes responsibility—and Zewail feels it strongly. That's why he cannot bring himself to summarily and carelessly brush off throngs of Egyptians calling for him to become president, even though he has no desire to hold that office. "In this historic moment, when people have died for such an important cause and Egyptians are asking me to help

---

Margaret Warner of "PBS NewsHour" recently likened interviewing Zewail in downtown Cairo to "sitting in public with Bono and Einstein combined."

---

by becoming president, it's my duty, at the very least, to think about it," he says. But until now, that's as far as it has gone.

That same sense of duty and a desire to help people better themselves are what drive Zewail to continue globe-trotting, especially to Muslim-majority countries, where he promotes education reform and investment in science. He knows he has given "the man and woman on the street" hope that those improvements will come, he says, because they tell him of their newfound optimism and shared desire to strengthen their countries' educational systems. Their governments, however, have thus far been less responsive to Zewail's encouragement and have remained reluctant to substantially increase allocations for education and science.

Even before becoming a U.S. science envoy, Zewail was busy trying to raise awareness of the importance of obtaining a rigorous education and establishing solid science and technology bases in developing countries. In lectures and in written opinion pieces, for example, he called upon developed nations not only to provide financial aid for these causes, but to become partners with the recipients and provide them with expertise, educational guidance, and implementation plans (*Nature*, DOI: 10.1038/35071136). He also helped bring prestige to highly committed young people by setting aside part of his Nobel Prize winnings to establish three prizes. Two are awarded annually by the American University in Cairo for outstanding academic performance in science and service to society. A third one is presented at the Cairo Opera House each year for achievements and creativity in the arts.

LONG BEFORE any of those efforts in education took shape, Zewail was shaping his own education. Had this year's Priestley Medalist been honored for "developing revolutionary methods for the study of ultrafast processes in chemistry, biology, and materials science" in any year other than this one—a year in which a key chapter in the history of his native Egypt and the Middle East is being rewritten—this biographical sketch might well have started

with the little sign posted by the medalist's parents on the door of young Zewail's bedroom naming the serious schoolboy who slept and studied there "Dr. Ahmed." For even when Zewail was a little kid, his parents recognized and encouraged their son's inquisitiveness and passion for learning. "If I got a 98 out of 100 on a test," Zewail recalls fondly, "my father would scold me playfully, '*Ya-bni,* my son, what about the other two points?'"

CRUISING THE NILE This 1999 photo captures the Zewail family, Nabeel (in his father's arms), Hani (sitting with his mother, Dema), and Zewail daughters, Amani (left) and Maha, boating down Egypt's famous river.

And so it was that Zewail's drive to earn top scores in school continued throughout his younger years in his hometown, Disuq, and in nearby Alexandria, where he went on to attend university. There, as a science student majoring in chemistry, Zewail's academic performance distinguished him as the highest-ranked student in his class and qualified him to serve thereafter as a *mu'id,* or demonstrator, equivalent to a teaching or research assistant. That position, which just seven chemistry students in a class of roughly 500 managed to attain, was a big deal because if a *mu'id* earned a Ph.D., the university would offer him or her a position as a professor.

Eager to pursue a career in academia, Zewail completed his master's study at the University of Alexandria and applied to U.S. schools to do Ph.D. work. He was accepted to the University of Pennsylvania, and after clearing countless hoops and hurdles imposed by Egyptian bureaucracy, he finally set off for the U.S. in 1969. He had little cash in hand and equally little command of the English language, and he soon found that he was in for one giant culture shock.

Although he had passed all of his courses with flying colors, Zewail's science education in Egypt was very traditional and didn't cover much quantum mechanics, group theory, or other topics in modern chemistry and physics needed for the foray into spectroscopy on which he was embarking. Likewise, he had had little experience with lasers and advanced instrumentation up to that time. But learning science was his forte, and in no time he came up to speed. Acclimating to American culture was an entirely different story.

Picture a 20-something, well-groomed, conservative Arab man in a starched white shirt, pressed dress pants, and coat and tie walking around the campus of a U.S. university in the free-spirited days of 1969. Surrounded by fellow grad students sporting colorful T-shirts and jeans with holes, Zewail wasn't quite sure what to make of the curious dress habits of the young people in his strange new world.

> ## "Ahmed's work has changed not only what we know about chemical reactions, but what we believe is possible to know."

MITCH JACOBY/C&EN

The refined *mu'id* from the banks of the mighty Nile River was even less sure what to make of their social habits. He smiles now when he recalls how a young man and woman, students in the first lab course he taught, decided right in the middle of running a titration that the time was ideal to start making out. "These two students began kissing, right in front of me, right in front of the whole lab, with passion," Zewail emphasizes laughingly. "In Egypt, this scene would have been impossible."

THE SCIENTIFIC and cultural education Zewail acquired at Penn and later at the University of California, Berkeley, where he conducted postdoctoral research, served him well. Although it would be quite a while before he tried on a pair of jeans—and even then only ones without holes—and a while more before he would come to understand why a man came running past his Berkeley lab one night wearing a mask and nothing else ("streaking" doesn't translate easily into Arabic), Zewail's easy manner and charm soon brought many close and long-lasting friendships among people of many nationalities and cultures. And his gift for devising insightful and probing experiments, and for explaining their meaning in simple, straightforward language, soon earned him a reputation as a topnotch scientist.

Within a few years of taking a faculty

ULTRAFAST MICROSCOPY
Zewail's current research efforts are aimed at advancing ultrafast imaging in biology and materials science via laser-driven electron microscopy methods pioneered by his group.

position at Caltech in 1976, Zewail and his new research group were thoroughly immersed in pushing the limits of molecular dynamics. The team soon developed femtosecond laser methods for exciting sample molecules with a "pump" beam and then quickly probing them with a second light pulse. By carefully tuning the interval between the two pulses, the researchers recorded series of snapshots that captured reactants evolving to products by way of transition states that existed for mere quadrillionths of a second.

That body of work, for which Zewail was honored with the Nobel Prize, "has made it possible for us to observe and understand some of the most intimate details of the processes by which substances are transformed," says Zewail's Caltech colleague David A. Tirrell. "Ahmed's work has changed not only what we know about chemical reactions, but what we believe is possible to know," Tirrell adds.

Years after those pioneering experiments were conducted, Harvard University's George M. Whitesides still finds it "truly remarkable" that Zewail's team devised ways to observe dynamics of molecular bonds on a timescale several orders of magnitude shorter than the time required for a bond to vibrate.

Thinking back to the days when some of those key experiments were being carried out, Zewail recalls the period as being "filled with thrilling moments," adding

that the excitement meant he wasn't sleeping much at that time. Neither were his students.

Jennifer Herek, who joined the Zewail group in 1990 as a graduate student, says the experience was "intense." Now a physics professor at the University of Twente, in the Netherlands, Herek says Zewail "always was so enthusiastic, and motivated us to work hard and long hours." She recalls the way Zewail engendered a "collective spirit" that inspired the group to get things done. "He often told us we would look back on this time as the best days of our lives." At the time she wasn't so sure. "But he was absolutely right," she acknowledges.

Working in the Zewail group at around the same time as Herek did, Dongping Zhong, now a physics professor at Ohio State University, says that one of the most important lessons he learned from Zewail is that "in order to succeed in science, you have to be passionate about your work." But that was easy, he adds, because Zewail's passion for science was "highly contagious."

WINNING A NOBEL PRIZE in 1999 didn't temper Zewail's drive to push the frontiers of science. It was around that time that his group was developing ultrafast electron diffraction techniques, which led to the more recent development of four-dimensional electron microscopy. That technique enables the temporal and spatial behavior of matter to be studied directly and simultaneously as events unfold in real time. These latest efforts are spawning new imaging methods that experts say may revolutionize biological and materials sciences because they can provide 3-D views of nanometer-scale objects evolving on the femtosecond timescale (C&EN, June 28, 2010, page 11).

And just as the fame and never-ending list of obligations that come with international scientific recognition didn't dampen Zewail's drive to advance science, they also didn't diminish his care and concern for the welfare of the people in his group. Students and postdocs whose era in Zewail's group spans from more than 20 years ago to as recently as last year, unanimously agree that their adviser truly cares about them.

For a man like Zewail, who has shown himself to be a concerned citizen of the world, that characterization hardly comes as a surprise. "Ahmed is among the most decent of people," says Harvard's Whitesides. He adds, "He's a warm and dignified man." ∎

DREAMING THE FUTURE

I AM HONORED and gratified to receive the Priestley Medal. This highest honor of the American Chemical Society comes from a society I have been associated with for decades and with which I continue to have strong relations, not only as a member and fellow, but also with its institutions, the board of directors, the society journals, and the super-dynamic Executive Director & CEO Madeleine Jacobs. Recently, Madeleine asked me to preside over the 44th International Chemistry Olympiad, and as many of you know, when Madeleine calls you with her typical affection and enthusiasm, you simply cannot say no!

When ACS was established in 1876, its founders were luckily unaware of, or perhaps chose to ignore, the words of the sage Thomas Jefferson, who in 1809 wrote in a letter to his nephew, "If you are obliged to neglect any thing, let it be your chemistry. It is the least useful and the least amusing to a country gentleman of all the ordinary branches of science."

Jefferson went on to promote the virtues of farming over chemistry! Fortunately, many people have not shared Jefferson's preference for farming, including a certain graduate of the Oregon Agricultural College by the name of Linus Pauling. Linus famously said, "Chemistry is wonderful! I feel sorry for people who don't know anything about chemistry. They are missing an important part of life, an important source of happiness, satisfying one's intellectual curiosity." Pauling received the Priestley Medal at the age of 83, so make sure to live long!

For all awards, I believe the personal satisfaction one feels in receiving them comes from the recognition by one's peers and from the history of the award itself. The medal I am receiving carries the name of Joseph Priestley, a great figure of the 18th century who achieved scientific immortality for his discovery of oxygen. As important, Joseph Priestley was also a minister who fought for educational reform and personal liberty at a difficult time when Europe was infected by religious fanaticism. In 1794, he emigrated from England

PRIESTLEY MEDAL ADDRESS

AHMED H. ZEWAIL

to America, where he became a friend of Jefferson, who sought his advice on plans for founding the University of Virginia. Priestley's move to America is telling of the great opportunity this country has offered to immigrants, including myself.

Following the ACS announcement early last year, I received a large number of congratulatory notes from friends and colleagues around the world, but the scientific contributions cited for the award would not have been made without the dedication of a large number of research scientists, postdoctoral fellows, graduate students, administrators, and staff. To all of them, I am grateful.

I would also like to take this opportunity to salute my colleagues who are being hon-

ored by ACS awards, and especially my former postdoctoral mentor at the University of California, Berkeley, Charles B. Harris, who just received a very strangely named award, the ACS Ahmed Zewail Award in Ultrafast Science & Technology. Last, but not least, are members of my family; to them and to the memory of my parents I dedicate the achievements being honored.

Traditionally, the Priestley addresses map trajectories of the past or, as in the case of my friend George Whitesides, give a futuristic outlook on chemistry. Tonight, I will use my own voyage from Egypt to America to reflect on the odyssey's lessons for an immigrant in search of "making a dream." I will then discuss the challenges facing this country and the world, and the role science can play in diplomacy.

MAKING DREAMS. At the age of 16, I was among the millions around the world who were astounded by a dream in the making. On Sept. 12, 1962, at Rice Stadium, President John F. Kennedy said, "We choose to go to the moon. We choose to go to the moon in this decade and do other things, not because they are easy, but because they are hard, because that goal will serve to organize and measure the best of our energies and skills, because that challenge is one that we are willing to accept, one we are unwilling to postpone, and one which we intend to win."

These historic words changed the course of space exploration and much else as well. In a thousand years from now, American civilization will be admired for, in addition to its founding Constitution, the 1969 landing on the moon and the beginning of space exploration, just as we still pay tribute to ancient Athenian democracy and marvel at the pyramids of Giza.

President Kennedy was in the right place and the right time to set forth this audacious vision for a nation. Similarly, for individuals, dreams are defined by one's vision, from intuition or insight, and by the place and time where one lives.

In my case, my parents must have anticipated my thirst for knowledge and decided on a birth in a town flanked by two cities of knowledge, Alexandria and Rosetta. Alexandria, where I went to college, is a place steeped in history with its ancient library and as the intellectual home of great scientists, such as Euclid, Archimedes, and Hypatia. Rosetta is where the famous stone was discovered, with its three engraved scripts: ancient hieroglyphs, Demotic, and ancient Greek. Despite these auspicious surroundings and the commendable education I received in Egypt, my dreams were modest.

EVOLUTION OF DREAMS. My first dream at the age of 15 was physically dynamic. I was curious about how a solid—wood—produces a gas that can be lit with a match. To investigate, I built a glassware apparatus in my bedroom and carried out an experiment to observe the burning of wood. My apparatus collected the resulting gas, which I duly lit with a match. This activity, of course, made my mother very unhappy.

But I did not worry about such details because I was driven by the curiosity of a schoolboy's mind. I was not thinking of Michael Faraday's search for the origin of combustion in his famous lecture, "Chemical History of a Candle." Neither was I aware of Lavoisier's use of a "burning glass" on various substances or even Priestley's discovery of oxygen, although I knew that oxygen existed.

I was merely following my intuition, which led me along the simple path of discovery: I must have been intrigued by a natural phenomenon, "light from combustion," and I then asked a simple question, "Why?" and designed a direct experiment. I recall explaining the results to my fellow pupils, and I think I convinced many of them to see the beauty of simplicity in a scientific inquiry. Beautiful experiments and observations often appear "trivial" in retrospect, but their findings are usually of a fundamental nature.

Such beliefs have been with me all along and so has my curiosity about "changes of matter," or in today's language, "the dynamics of matter's transformation," perhaps because of what I learned early on—

"ALMOND BLOSSOM" What do this painting by Vincent van Gogh and the nature of scientific discovery have in common? To me, the beauty of the big picture and the unpredictability of its details.

that chemistry, as a science, had its roots in ancient Egypt. Some historians believe that the word "alchemy," from which "chemistry" is derived, is a corruption of the word *khemia* or *keme*, which literally means "the black land," an ancient name for the land along the banks of the Nile, where the flooding river would transform the color of the soil during inundation.

Coming to the U.S. in 1969 was certainly the gateway to bigger dreams. In this post-*Sputnik* era, America was second to none in the opportunities it offered, especially to young people. Besides the rich and free culture, there was the feeling that the sky was the limit. Both of my mentors, Robin Hochstrasser at the University of Pennsylvania and Harris at UC Berkeley, instilled in me the significance of fundamental science, and I had the feeling that funding was not an issue for them. They spent most of their time as scientists and not as managers of science. Through this experience, I also saw the difference in so-called small and big science, especially when I became an IBM Fellow working at the fantastic fa-

---

Chemistry can and should remain a fundamental science, providing new tools and defining new concepts, but with its lens focused on significant questions in emerging areas of complexity, from nanoscience to physical biology.

cilities of Lawrence Berkeley National Laboratory.

While at UC Berkeley, I received an offer from California Institute of Technology. Soon, as a faculty member at Caltech, I found I was once again in the right place at the right time. The legacy of Pauling was all around, with research focused on the structure of molecules both at Caltech and elsewhere. I was in a unique position to begin research on dynamics of molecules. At the time, the concept of coherence in chemistry was foreign, and some renowned chemists, including three Nobel Laureates, did not appreciate its significance and did not see much value in it. But Caltech gave me the opportunity to be intellectually independent with plenty of room to pursue my goals. I worked with an exceptional research group, and in 1987 we published the first paper on femtochemistry, reporting on the probing, and potentially the controlling, of the coherent motion of atoms in reactions and during their ephemeral transition states. Some scientists were concerned about issues such as the uncertainty principle and the general applicability of the approach, but I have seen that after 1999 they became convinced!

In this regard, I must mention my dear friend and great supporter, the late Richard Bernstein. Dick's unqualified enthusiasm for the development of femtochemistry, the mentoring days we spent together at Caltech when he was on sabbatical there as a Sherman Fairchild Distinguished Scholar, and the joint papers we wrote, including a *Chemical & Engineering News* feature article, are experiences that I treasure. Dick predicted femtochemistry would be recognized by the Nobel Prize, but sadly he died without witnessing his prophecy come true.

The Nobel Prize is a great honor, but it comes with one expectation—stop science and become an expert (or at least a pundit) on everything from the stock market to the future of humanity. People found it hard to take my desire to continue research seriously.

I had a bigger dream than femtochemistry, namely, to be able to visualize matter's transformations, not only in time but also in space—to chart the movements of atoms in all four dimensions. Over the past 10 years at Caltech, we were able to develop 4-D electron microscopy to do just that. We now use ultrashort packets of electrons to probe systems with subnanometer spatial resolution and femtosecond time resolution. As well as opening up these extraordinary space-time dimensions to basic research, this realization of our big dream has myriad potential applications in chemical, biological, and materials sciences. None of this would have been possible without being in the right place, Caltech, at the right time, with the funding from the Gordon & Betty Moore Foundation.

In this more recent endeavor, another dear friend, Sir John Meurig Thomas of Cambridge University, became an enthusiastic supporter of the new development. He appreciated early on, when some skepticism arose, the significance of 4-D real space and diffraction imaging. John is a world expert on 2-D and 3-D electron microscopy, and together in Pasadena we wrote a monograph, published two years ago, entitled "4D Electron Microscopy: Imaging in Space and Time" (Imperial College Press, 2009). Over the years, I have enjoyed his poetic perspectives on science and scientists.

CHEMISTRY DREAMS. So far, I have spoken about dreams in areas of particular interest to me—but what about the discipline of chemistry at large? Some in the profession think that chemistry in the 21st century is now at "the end," perhaps only useful in service to other fields. Even more broadly, some writers of popular books have claimed the end of science! These views are shortsighted, to say the least. I believe that the opportunities for basic research in chemistry this century are more exciting than ever, provided that we do not restrict our vision to orthodox boundaries.

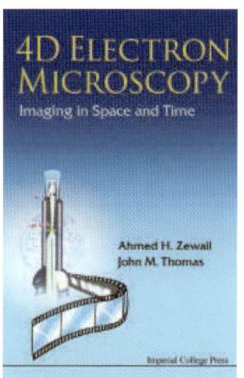

POST-NOBEL SCIENCE This 2009 book written by Thomas and myself describes the new field established over the past decade since my receipt of the 1999 Nobel Prize in Chemistry.

Breakthroughs will continue to emerge when we understand how and why systems of thousands of atoms in macromolecules, materials, and cells function coherently and as if with a directed purpose. In the end, we may or may not find that the whole is greater than the sum of its parts and learn how complex systems in nature produce unique behaviors describable by classical mechanics even though they are made from the probabilistic quantum world of atoms and molecules. If we resolve these profound mysteries, it will affect numerous areas of materials and life sciences, and even physics.

From my perspective, chemistry will become merely a service to other fields only if we lose sight of its primary objective. Our field can and should remain a fundamental science, providing new tools and defining new concepts, but with its lens focused on significant questions in emerging areas of complexity, from nanoscience to physical biology. The cause is helped if more champions articulate the beauty of chemistry's fundamentals and the big picture of its mission, globally and in the U.S. We should look ahead, unswayed by pressures of fads or funding, and be driven by our own curiosity.

CURIOSITY-DRIVEN DREAMS. As brilliantly conveyed by Lewis Carroll in "Alice's Adventures in Wonderland," curiosity is the key to explorations that go beyond the "known unknowns" to delve into the "unknown unknowns." Recently when I was on an official visit to Southeast Asia, a prime minister asked me, "What does it take to get a Nobel Prize?" I answered im-

ALICE'S KEY Curiosity is the key to explorations that go beyond the "known unknowns" to delve into the "unknown unknowns."

mediately: "Invest in basic research and recruit the best minds." This curiosity-driven approach seems increasingly old-fashioned and underappreciated in our modern age of science. Some believe more can be achieved through tightly managed research—as if discoveries are predictable. This attitude is an unfortunate misconception that affects and infects research funding.

There are countless examples of valuable breakthroughs that came from research driven by the curiosity of individuals. Perhaps the best example comes from the work done to develop the maser and the laser by Charlie Townes and colleagues. Last summer in Paris at a celebration of the 50th anniversary of the laser's invention, Townes reminded us that he was driven at the start only by an intellectual curiosity about the microwave spectroscopy of molecules, which later led to the invention of the ammonia maser. In this odyssey, fundamental issues had to be addressed: how to enhance Einstein's stimulated emission over absorption, how to sustain the gain, and how to "beat" the uncertainty principle and achieve monochromaticity through coherence. The laser was called "a tool in search of a problem." No one imagined the technological impact it has today, from eye surgery to the information technology revolution.

As I mentioned at the Paris celebration, in my own career it was curiosity about questions pertinent to coherence and the uncertainty principle that brought about my group's contributions first to femtosecond science and now 4-D electron microscopy. I doubt if the first grant proposal I wrote about coherence in dynamics, which had no immediate relevance to anything with so-called broader impact, would be funded today.

Quantum mechanics, relativity, the expanding universe, and the deciphering of the genetic code

U.S. SOFT POWER
Respondents to a recent poll by Pew Research Center involving 43 countries overwhelmingly said that what they most admire about the U.S. is its leadership in science and technology. The much-praised American entertainment industry came in a distant second.

were discoveries made down the rabbit hole of curiosity. So, too, are revolutionary technologies such as MRI (developed from curiosity-driven research about the spin of an electron) and the transistor (discovered as a result of curiosity about the properties of electrons in semiconductors). The industries that followed now constitute the backbone of worldwide communications and the global economy. Curiosity pays! As Francis Bacon said, "A wise man [woman] will make more opportunities than he [she] finds."

GLOBAL DREAMS. Curiosity-driven science and technology has paid off for America, not only with wealth and overt power, but also with soft power, the power that sways hearts and minds. It would be wise and timely to make use of this soft power in global affairs. In today's world, America's soft power is commonly thought to reside in the popularity of Hollywood movies, Coca-Cola, McDonald's, and Starbucks, but studies tell a different story. In a recent Pew Research Center poll involving 43 countries, 79% of respondents said that what they most admire about the U.S. is its leadership in science and technology. The artifacts of the American entertainment industry came in a distant second.

What I found unique to this country in the 1970s as a foreign student is what much of the world continues to value most about the U.S. today, namely, its open intellectual culture, its great universities, its capacity for discovery and innovation, and its spirit of entrepreneurship. The U.S., by harnessing the soft power of science in the service of diplomacy, can bring the best of its cultural heritage to bear on building better and broader relations with the world at large. In many ways, science embodies the core values of what the American Founders called "the rights of man" as set forth in the Bill of Rights: freedom of thought and speech and commitment to equality of opportunity.

BACK TO THE FUTURE. In his 2009 Cairo speech, President Barack Obama articulated a new initiative for cooperation and partnership that emphasizes the role of science in diplomacy, particularly with Muslim-majority

PREDICTION MAP This rendering depicts the 1999 vision I envisaged for the future. Femtochemistry is branching like a mighty river to physical, organic, and other areas of chemistry, touching single-molecule-, attosecond-, and angstrom-resolution measurements along the way. The culmination is in 4-D imaging, control, and biodynamics. Much of this science has already been witnessed in labs around the world.

countries. Shortly afterward, I was appointed the U.S. science envoy to the Middle East, and I embarked on a diplomatic mission that took me back to where I came from, but now with a different dream. What I learned from touring and seeing the state of science and education in the region was cause for some alarm, but also for considerable optimism.

Education in the U.S. faces many problems. The majority of the countries I visited face similar difficulties, but they also confront much more severe troubles that impede national and global progress. Yet there are positive signs as well. Countries such as Malaysia, Turkey, and Qatar are making significant strides in education and in technical and economic development. Egypt, Iraq, Syria, Lebanon, Morocco, and Indonesia are examples of countries rich with talents—about one-third or more of their populations are under the age of 30. We should not forget that the history of human civilization began and flourished in the Middle East. Today, there are many people from these countries living in the West who have excelled in all fields of endeavor. The latent capability of the people in the Middle East and in Muslim-majority countries elsewhere lies undiminished until circumstances are suitable for its development.

I recently read an important study that left me in awe of the knowledge demo-

graphics of our planet. In "Educating All Children: A Global Agenda," Joel Cohen and David Bloom argue that the aim of achieving primary and secondary schooling for all children is urgent and feasible, and yet more than 300 million children will not be in school in the year 2015. Every effort should be made to change this bleak picture so we may hope for a better future for our world. The soft power of science has the potential to reshape global diplomacy and at significantly lower expense than that needed for the hard power of military involvement.

EPILOGUE. I would like to end by stressing the virtues of dreaming the future. Dreams evolve dynamically through space and time. Being in the right place at the

right time can be a matter of luck, but dreamers must also actively seek out opportunity. Dreamers must be willing, and allowed, to take risks. In our profession of scientific exploration, as in the arts, the most creative work will materialize when intellectual curiosity is unhindered by the forces of bureaucracy and weighty management. As Louis Pasteur said, "Chance favors the prepared mind," but without the appropriate milieu, a dream cannot materialize.

This country was established as a dream, explored outer space propelled by a vision, and pursued a dream of a science policy—the "endless frontier" in the words of Vannevar Bush, after World War II. Despite current problems, the U.S. continues to lead the world through the power of knowledge. In the 21st century, America must return to its guiding principles and defining characteristics. I am hopeful that we will chart a new policy for innovation that is inclusive of international science diplomacy for partnerships in development. Some may argue that it is naïve to think of such idealistic values in our imperfect world, but the influence of science diplomacy is in the best interest of the U.S. Through the power of knowledge and curiosity, we can efface ignorance and shape a future that binds cultures and civilizations.

In his Stockholm address, the winner of the 1988 Nobel Prize in Literature, Naguib Mahfouz, reminded us of our responsibility as citizens of a world made of haves and have-nots. He said, "In this decisive moment in the history of civilization it is inconceivable and unacceptable that the moans of mankind should die out in the void. ... Today, the greatness of a civilized leader ought to be measured by the universality of his [her] vision and his [her] sense of responsibility towards all humankind. The developed world and the third world are but one family." ∎

"A wise man [woman] will make more opportunities than he [she] finds."
—Francis Bacon

SHUTTLE DIPLOMACY During the Egyptian revolution last month, I (back row, center) met in Cairo with the leaders of young demonstrators, activists, and organizers with the purpose of carrying their views to Egypt's top brass.

Revolutionary Dreams

This address was written before the Egyptian revolution on Jan. 25 of this year. The revolution gave birth to a dream that materialized on this historic day, and I had the privilege of witnessing the event unfolding in real time with millions of Egyptians peacefully uprising for democracy. Living through such a dream is the experience of a lifetime; in this case, not only for me, but also for 85 million Egyptian citizens. Those who died in the struggle were seeking a better future. I hope we can honor their sacrifice by fostering education and science to help forge the future they fought for.

Proceedings

of the

American Philosophical Society

Held at Philadelphia for Promoting Useful Knowledge

VOLUME 150 • NUMBER 4 • DECEMBER 2006

THE AMERICAN PHILOSOPHICAL SOCIETY

INDEPENDENCE SQUARE: PHILADELPHIA

2006

Franklin's Vision[1]

AHMED H. ZEWAIL

Linus Pauling Chair Professor and Director of the
National Science Foundation Laboratory for Molecular Sciences
and the Physical Biology Center for Ultrafast Science and Technology
California Institute of Technology

O N THIS SPECIAL OCCASION of the Tercentenary, I am especially delighted to speak in honor of a polymath and an American icon, Benjamin Franklin (fig. 1). Since his death in 1790, Franklin has been revered, memorialized, and made into an educational, financial, and political icon. Through his collective work this sage has climbed to the apex of human endeavor in the sciences, public service, and statesmanship in international relations. Such great heights for a man of wit and wisdom are reached by very few in the world, both then and now.

I have real connections to Franklin, though not biological in nature. My first science home in America, the University of Pennsylvania, was founded by him; the pre-Nobel recognition I received from the Franklin Institute was the medal in his name; and election to this august society has indeed strengthened my bonds to Franklin's home of knowledge and to Franklinian ideals of "promoting useful knowledge." In my office I have his bust for a daily reminder of what it means to be a scientist in service of society and a citizen of the world at large.

For me personally, Franklin is a hero, not only for his unique and remarkable scientific contributions in the 1700s, but also for his humanitarian vision and his belief in the power of learning. He best used his own power as an accomplished scientist to influence world politics and peace. Perhaps the greatest of all of his achievements was his effort to secure America's independence and peace with England. Today, it is Franklin's vision, with his spirit of compromise and eloquence, that we need in order to reach a dialogue and peace in our troubled world.

Some attribute the origin of calamity and conflict to a clash of civilizations. I do not. Being a cultural product of both "East" and "West"

---

[1] Read 27 April 2006, during the Annual General Meeting celebrating the Franklin Tercentenary.

FIGURE 1. *Benjamin Franklin.* Early nineteenth-century copy after Joseph Siffred Duplessis.

with a voyage (1) of confluence, not clash, I do not find a fundamental basis for the so-called "clash of civilizations." What is important is to emphasize cooperation, not confrontation, and to understand that we live in an interdependent "flat world" (in the words of Thomas Friedman) that cannot be peacefully sustained with huge disparities in wealth and conspicuously inconsistent policies (2, 3). Let me quote what Franklin said more than two centuries ago in *Poor Richard's Almanac*:

> Who is wise? He that learns from everyone.
> Who is powerful? He that governs his passion.
> Who is rich? He that is content.

These words radiate vision, thought, and humor. Franklin added, "Who is that? Nobody." True, perhaps, but an important point is that being rich and powerful has meaning and responsibility, and that hegemony, if we are wise and learn from history, does not work in the end.

But this is not the "Franklin's vision" that I will be discussing in the remaining time. Rather, I would like to ask how, at the atomic and molecular level, did Franklin actually *see*?

Vision is the result of the conversion of light energy to an electro-chemical impulse (fig. 2). The impulse is transmitted through neurons

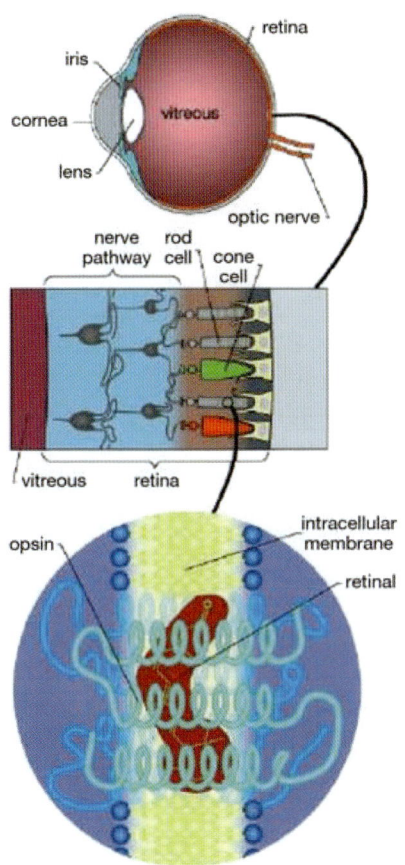

to the brain, where signals from all the visual receptors are interpreted. One of the initial receptors is a pigment called *rhodopsin*, which is located in the rods of the retina. The pigment consists of an organic molecule, *retinal*, in association with a protein named *opsin*. A change in shape of retinal, which involves the twisting of a chemical double bond, apparently gives the signal to opsin to undergo a sequence of dark (thermal) reactions involved in triggering neural excitations.

How does the complex system of vision render such a selective function? The speed of twisting is awesome. The primary motion occurs in 200 femtoseconds (one femtosecond is a millionth of a billionth of a second; a femtosecond to a second is what a second is to 32 million years). Such rapidity indicates that the light energy is not first absorbed, then redistributed to eventually find the reaction path of twisting. Instead, the entire process proceeds in a coherent manner; that is, a packet of waves in the language of quantum physics.

FIGURE 2. Molecular basis of vision

This coherence is credited with making possible the high (70 percent or more) efficiency of the initial step, despite the large size of the rhodopsin molecule and the many possible channels for dissipation of energy. That is what gave Franklin healthy vision, and enables you and me to see even when ambient light is dimmed.

However, we still do not understand why such complex systems with thousands of atoms, having 3n-6 degrees of freedom (vibrational motions), where n is the number of atoms, have the selectivity for the entire motion to be along one (global) path, and how they fold into a unique structure. In fact, this represents a paradox that is central to molecular diseases. How does a protein acquire its native conformation? We might think that the protein will try random visits of all possible conformations until it eventually reaches the correct conformation. How long would it take to do so? For a protein with N residues, each with the ability to twist or contort, there exist about 10^N possible conformations (ignoring other possible motions involving side chains). With 100 residues, the protein will need time that far exceeds the age of the universe (12 billion years)—obviously, if that were true, we would not be here!

This Levinthal's paradox can be overcome only if biosystems on the landscape of their motion are able to select with guided bias the native conformational structure, and in reasonable time, much shorter than a second. If, in the process, a protein does not follow this path of folding, i.e., misfolds, then we are in trouble, as a biopolymer protein becomes abiological—a chemical polymer—and we contract diseases such as Alzheimer's, a neurodegenerative brain disease. The deposition of proteins in the form of amyloid fibrils or plaques as a result of misfolding and aggregation is believed to be associated with other diseases such as Parkinson's, type 2 diabetes, and also other amyloidosis. The deposits, depending on the disease, can be in the brain, in skeletal tissue, or in other organs.

To understand the cause of the macromolecular change, we must find new ways to visualize the structures and the motion of atoms on the so-called energy landscape. With the discoveries of X-rays in 1895, and of the electron in 1897, it became possible to determine the static and invisible architecture of molecules in three dimensions (3D). The complexity of structures determined has increased over the years, from two-atom table salt (sodium chloride) to DNA, proteins, and complexes such as biomotors, ion channels, the ribosome (the "factory" that produces proteins), and polymerases (the machines that transcribe the genetic information). Similarly, major advances have been made in resolving the dynamics.

In over a century of developments, the time resolution has now, with the invention of lasers in 1960, reached the time scale (femtosecond) for atomic motions. Applications in all phases of matter and for biological systems have been the subject of study in the last two decades; the field was dubbed femtochemistry; similarly, femtobiology for related biological studies. As this ultrashort time scale was reached, there were conceptual and technical barriers to be crossed, and dogmas to be changed. Conceptually it was realized (4) that neither the uncertainty limit of time and energy nor the numerous degrees of freedom of complex systems would represent a fundamental limitation. What is important is to prepare and probe the atoms *coherently* and to ensure that the time of preparation is shorter than their de-coherence time, i.e., when the atoms are moving in unison. With femtosecond timing, the atom's motion becomes visible (1, 4).

Technically, we needed snapshots of atoms in motion. This stop-motion photography of atoms in molecules has historic roots, planted

a century after Franklin's death in 1790. In the nineteenth century, stop-motion pictures of animals were recorded for the first time, using light shutters and flashes. In France, Étienne-Jules Marey, a professor at the Collège de France, was working (1894) on a solution to the problem of action photography using *chronophotography*, a regularly timed sequence of images. Marey's idea was to use a single camera and a rotating slotted-disk shutter, with exposures on a single film plate or strip that was similar to modern motion picture photography. Marey applied his chrono-photographic apparatus in particular to humans and animals in motion, and to a subject that had puzzled people for many years: the righting of a cat as it falls so that it lands on its feet (fig. 3). How does the cat do it? Does its motion under gravity, without an external force, violate Newton's laws of mechanics? Does the cat have some special, magical physiology, or a command of some weird new physics?

By "slicing time" and freezing the motion during the fall, in the transition state of the righting, Marey was able to answer the question. First, the cat rotates the front of its body clockwise and the rear part counterclockwise, a motion that conserves energy

FIGURE 3. Motion of a cat, righting under the influence of gravity

and maintains the lack of spin. It then pulls in its legs, reverses the twist, and, with a little extension of the legs, it is prepared for a final landing. The cat instinctively knows how to move, as do high divers, dancers, and some other athletes. The answer to the puzzle was that the moving body was not rigid, and Newton's laws prevailed. However, scientists needed photographic evidence of the individual stopped-action steps to understand the mystery. At the time, these observations were thought-provoking; renowned scientists discussed in public their meaning and significance. J. Willard Gibbs (fig. 4) gave a talk on 4 December 1894 before the Mathematical Club at Yale with the title "On Motions by Which Falling Animals May Be Able to Fall on Their Feet." Marey's work and that of Eadweard Muybridge on galloping horses for the first time captured the behavior of animals (and humans) in motion.

For atoms in motion, we must consider their microscopic language (quantum mechanics) and dimensions. The contrast in *length* and *time* scales for the motion of the cat and the atom is awesome (fig. 5). For a definition of 1 cm, a cat speeding at 2 meters per second requires a time resolution of 0.005 second. But for a molecular

FIGURE 4. *Josiah Willard Gibbs*. From L. P. Wheeler, *Josiah Willard Gibbs: The History of a Great Mind* (New Haven, Conn.: Yale University Press, 1952).

structure in which atomic motions of a few angstroms (an angstrom, Å, is 10^{-8} cm) typically characterize chemical change, a detailed mapping of the motion will require a spatial resolution of less than 1 Å (about 0.1 Å). Therefore, the shutter time, or time resolution, required to observe with high definition atoms in motion at a speed of one kilometer per second (1,000 m/s) is 10 femtoseconds—a million million times shorter than what was needed for Marey's (or Muybridge's) stop-motion photography. On this time scale, for the most part, the quantum language can be replaced with the classical description of motion. Even though this development in the race against time (fig. 6) was recognized with the 1999 Nobel Prize, there remained the challenge of integrating both space (the 3D architecture) and time (the motion) in the visualization of complex systems.

Over the past five years progress has been made to define a new field of study (5), imaging in four dimensions (4D). The ability to visualize simultaneously both the structure and dynamics is fundamental to our understanding of the function in the realm of what is termed "complexity" and "emergence." The approach at Caltech involves the generation of

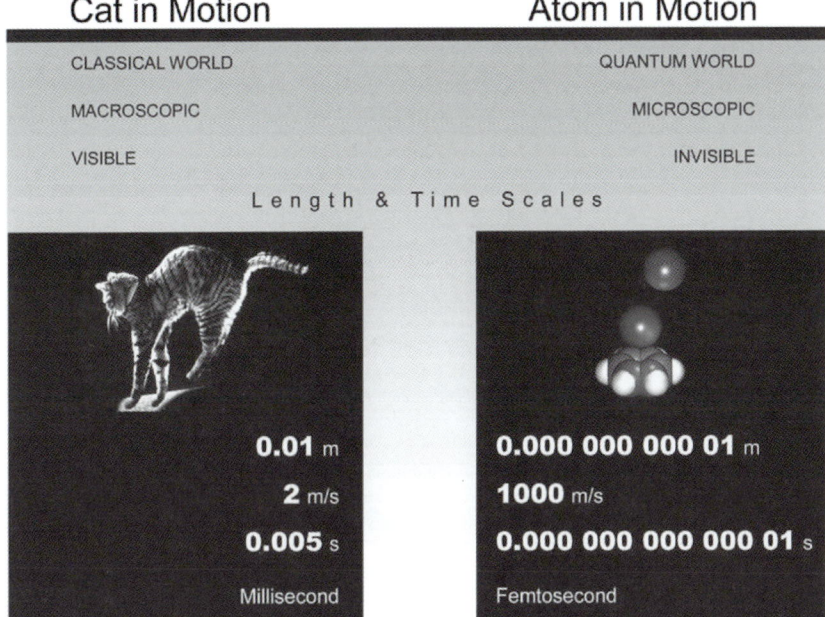

FIGURE 5.   Length and time scales for cats and atoms in motion

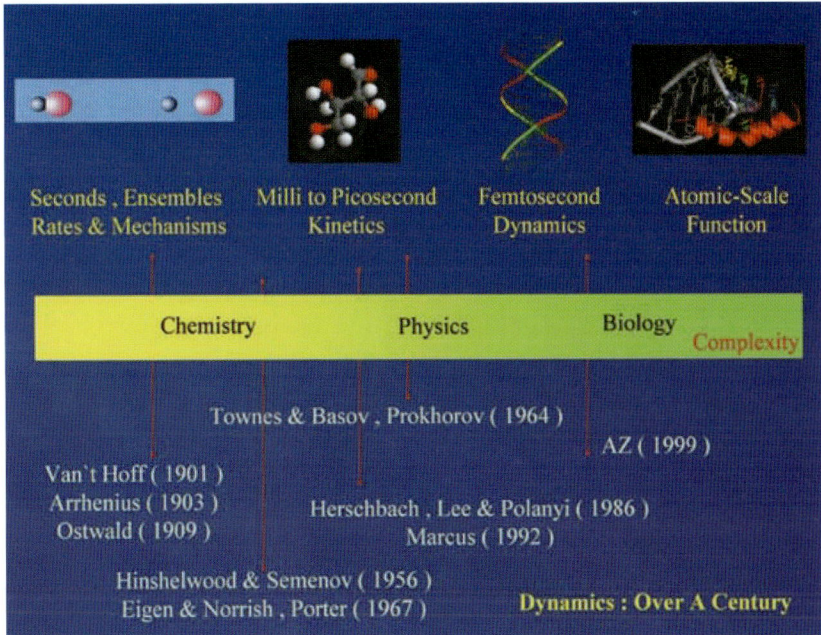

FIGURE 6.   The race against time in a century of developments, highlighted by
the Nobel Prizes awarded

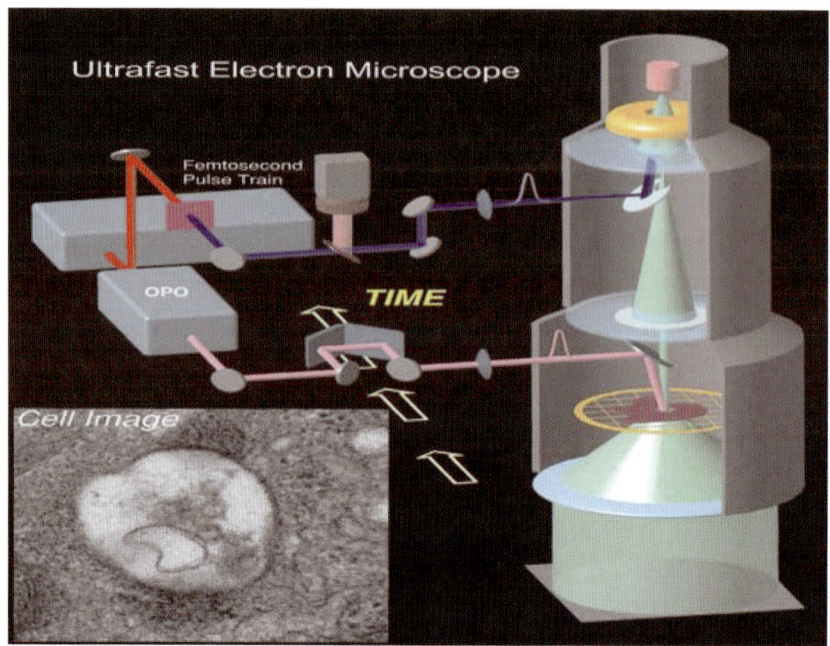

FIGURE 7.   4D microscopy and imaging of a biological cell

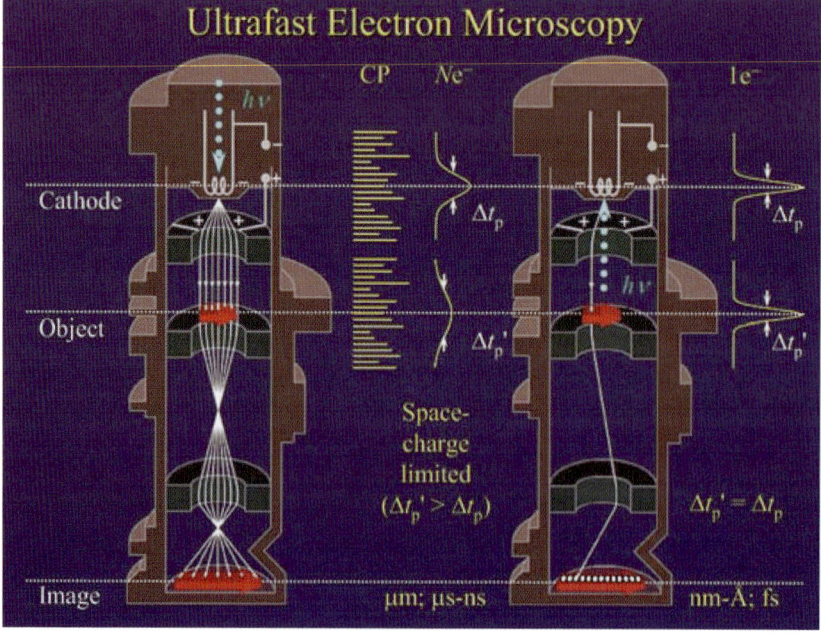

FIGURE 8.   Concept of single-electron microscopy

ultrashort (femtosecond and picosecond) *packets of electrons* (not light) for imaging of (and diffraction from) objects, molecular or cellular, with both spatial and temporal resolutions at the atomic scale. This 4D imaging, with time representing the fourth dimension, makes possible the study of the structural transformation of chemical reactions, materials, and biological systems (as discussed in the talk) (figs. 7, 8). The advances have been termed revolutionary by the previous speaker, Sir John Meurig Thomas,[2] because of the potential of imaging applications using timed, single-electron 4D microscopy and diffraction in different areas of science and technology.

One of the problems we studied recently (6) relates to an old question Franklin asked about "oil and water." In his experiments with oil and water one circumstance struck him with particular surprise. When a drop of oil is put on water, it spreads instantly and widely, becoming so thin as to produce prismatic colors and be nearly invisible. This problem is rich, as it exposes phenomena of a different nature. They form the basis of study of molecules and the forces between them at surfaces and in fields of chemistry and biology (especially cell membranes). The rapidity of the spread, the degree of molecular "liking" and orientation, and the stilling or calming of water with oil are issues that have been addressed over the past two centuries. What remained invisible is 4D structural dynamics of the molecules at the interface, which is nanometer in scale. Our study (6) focused on the microscopic extent of liking between water molecules and surfaces of varying forces of attraction, from being hydrophobic to hydrophilic in character. We also studied oil-like molecules to elucidate structural dynamics of self-assembled networks at interfaces (7). It is remarkable that Franklin's experiment, described succinctly in his letter to William Brownrigg on 7 November 1773, is still with us today! (Incidentally, in reference to this experiment, I asked my young son Hani, why don't vinegar and oil mix? He said he does not know, but they taste good!)

Visualization in space and time provides a myriad of opportunities at the interface of physics and chemistry with biology in what we have established as "physical biology" at Caltech. With the power of molecular and genetic tools integrated with 4D imaging, the focus is on the development of concepts for describing selectivity and global function. Physical biology as a new discipline aims at understanding mechanistically how physical forces and interactions govern biological function, from the molecular to the cellular scale. This focus on physical forces

---

[2] "A Revolution in Electron Microscopy," *Angw. Chem. Intl. Ed.* 44 (2005), 5563.

of biological structure-dynamics is distinct from the aim of mapping the engineering of information and its flow, the wiring of cells.

It remains to be seen whether *complexity*, involving the interactions among different components, and *emergence*, involving the selective function, require new theories of physics other than those we know about for atoms and molecules. What is certain is that the physical forces among these atoms and molecules result in a unique non-equilibrium state with continuous dynamic change. Life would not exist without the dynamic making and breaking (or twisting) of chemical bonds, and, remarkably, the forces important to the majority of functions of life molecules are weak, but dynamic. At the end, the function emerges coherently and through ambient fluctuations, and it involves many components.

Undoubtedly, as we progress in acquiring new understanding, new questions will emerge. Some of these questions are challenging. For example, how does well-defined classical function emerge from fluctuating quantum systems? What is the driving force for the bias directing the function? Does complexity make "the whole greater than the sum of its parts," and if so, why? On a different scale, we are back to a fifty-year-old question posed by Schrödinger—what is [the molecular basis of] life? It is hoped that in the coming fifty years, physical biology will endeavor to answer these and other questions through direct visualization (5, 8) of the elements of biology in space and time.

With its miracles and mysteries, time's resolution, direction (arrow), and meaning have occupied some of the best minds of science and philosophy (9, 10). As the fourth dimension, time is fundamental to all physical phenomena of our universe (we do not know what will happen as we reach the eleven dimensions of superstring or M-theory). Through time, we perceive what actually happens and what life is all about. Benjamin Franklin understood the importance of time and its centrality to our lives. Of time and life he said (*Poor Richard's Almanac*, 1746),

> Dost thou love life? Then do not squander time, for that's the stuff life is made of.

Franklin's vision has many dimensions that together formed the legacy of a great man in search of a better world. Although the biological molecules of Franklin's vision are now "dead," his life's work remains as a monumental book of knowledge. Maybe its cover has become food for worms, as he said about himself in his epitaph, but the contents and the work described shall continue to be a beacon of enlightenment. Forever his legacy will remain vivid in the pantheon of history.

Some Relevant References by the Author

1. 2002. *Voyage through time*. Cairo: American University in Cairo Press. Translated into French, German, Spanish, Chinese, Russian, Korean, and Arabic (twelve languages and editions).
2. 2001. Science for the have-nots. *Nature* (London) 410:741.
3. 2005. The future of our world. In *Einstein—Peace now! Visions and ideas*, ed. R. Braun and D. Krieger, 109. Weinheim: Wiley-VCH.
4. 2000. Femtochemistry. In *Les Prix Nobel 1999*, ed. T. Frängsmyr, 103. Stockholm: Almqvist, Wiksell.
5. 2006. 4D ultrafast electron diffraction, crystallography, and microscopy. *Ann Rev Phys Chem* 57:65.
6. 2004. With C. Y. Ruan, V. A. Lobastov, F. Vigliotti, and S. Chen. Ultrafast electron crystallography of interfacial water. *Science* (U.S.A.) 304:80.
7. 2005. With S. Chen and M. Seidel. Atomic-scale dynamical structures of fatty-acid bilayers observed by ultrafast electron crystallography. *Proc Natl Acad Sci* (U.S.A.) 102:8854.
8. 2006. With M. S. Grinolds, V. A. Lobastov, and J. Weissenrieder. Four-dimensional ultrafast electron microscopy of phase transitions. *Proc Natl Acad Sci* (U.S.A.) 103: 18427.
9. 2005. Time's mysteries and miracles: Consonance with physical and life sciences. In *Discovery to delivery*, ed. I. Serageldin and G. J. Persley, 9. Alexandria: Bibliotheca Alexandrina, and references therein.
10. 2001. The uncertainty paradox—the fog that was not. *Nature* (London) 412:279.

Discovery to Delivery

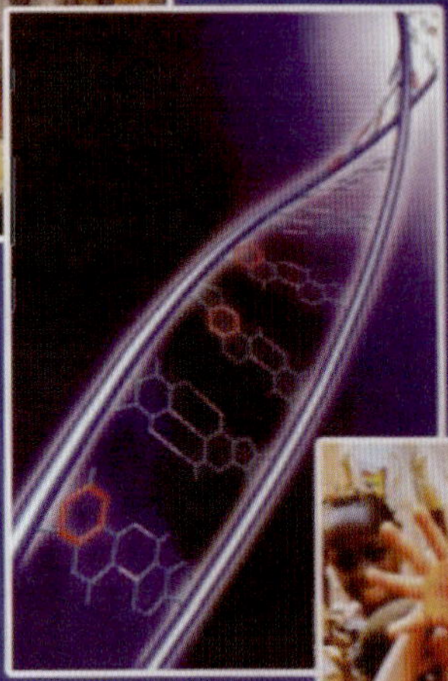

BioVision Alexandria 2004

Editors: I. Serageldin and G.J. Persley

BIBLIOTHECA ALEXANDRINA
مكتبة الإسكندرية

THE DOYLE FOUNDATION

Introduction

An Amazing Legacy

The very name of the Bibliotheca Alexandrina conjures up the image of a glorious past, of a shared heritage for all of humanity. For it was indeed at the Ancient Library of Alexandria that the greatest adventure of the human intellect was to unfold. Launched in 288 BCE by Ptolemy I (Soter) under the guidance of Demetrius of Phaleron, the temple to the muses, or Mouseion (in Greek), or *Museum* (in Latin) was part academy, part research center, and part library. The great thinkers of the age, scientists, mathematicians, poets from all civilizations came to study and exchange ideas.

They and many others were all members of that amazing community of scholars, which mapped the heavens, organized the calendar, established the foundations of science and pushed the boundaries of our knowledge. They opened up the cultures of the world, established a true dialogue of civilizations. For over six centuries the ancient Library of Alexandria epitomized the zenith of learning. The library completely disappeared over 1600 years ago…but it continues to inspire scientists and scholars everywhere. To this day, it symbolizes the noblest aspirations of the human mind, global ecumenism, and the greatest achievements of the intellect.

The Rebirth of the Bibliotheca Alexandrina

Sixteen-hundred years later, under the auspices of President Mohamed Hosni Mubarak, and with the continuous untiring support of Mrs. Suzanne Mubarak, it comes to life again. The Bibliotheca Alexandrina, the new Library of Alexandria, is dedicated to recapture the spirit of the original. It aspires to be:

- The World's window on Egypt;
- Egypt's window on the world;
- A leading institution of the digital age; and, above all
- A center for learning, tolerance, dialogue and understanding.

Editors

Foreword

The first Nobel Prize-winner in science from an Arab country, Dr. Ahmed H. Zewail is at the forefront of his field winning the 1999 Nobel Prize in Chemistry and establishing the whole new realm of femtochemistry. He is the Linus Pauling Chair Professor of Chemistry and Professor of Physics, and heads the NSF Laboratory for Molecular Sciences at the California Institute of Technology.

Dr. Zewail was born in 1946 in Damanhour, Egypt, 60 km south of Alexandria, where he showed early promise in his science classes; his family even jokingly put the sign "Dr. Ahmed" on his door. Dr. Zewail received his BSc and MS degrees from the University of Alexandria, where he was a brilliant student, giving "professorial lectures" at age 21. His advisors at the university recommended he do further training in the US, and against enormous odds (the 1967 war had just ended), he obtained a scholarship to University of Pennsylvania, from which he received his PhD. Upon completion of his doctorate, he went to the University of California at Berkeley as an IBM research fellow. In 1976, he accepted a position as assistant professor of chemical physics. Full professorship and tenure soon followed, and in 1990, he became the first Linus Pauling Chair at Caltech.

Much of the scientific endeavor of the 20th century was involved with the nature of time and the behavior of the smallest particles of matter. Einstein's theory of relativity ushered in a new era in physics, irrevocably linking time, matter, and energy, and subsequent research has attempted to close in on the actions of the tiniest particles in the smallest time-frames. Dr. Zewail's area of research, femtochemistry, is concerned with incredibly small fractions of a second, the femtosecond (10^{-15} of a second), and the problems of recording molecular and atomic movement at such minute time intervals. To explain just how small a femtosecond is, consider that from a femtosecond to one second is the same as from one second to 32 million years! Who can comprehend or even imagine 32 million years? Our history barely reaches seven to eight thousand years before receding into the mists of the very distant past. Yet that is the scale of the achievement of Dr. Zewail, experimentally, not just through the manipulation of abstract mathematical formulae.

Already at the University of Alexandria, Dr. Zewail had become interested in spectroscopy, an interest he pursued at the University of Pennsylvania, with special focus on molecular pairs. At Berkeley, he continued to investigate spectroscopy, this time with highly refined tools at his disposal. He worked with Dr. Charles Harris on making a picosecond laser (1 picosecond is 10^{-12} of a second, one-thousand times longer than 1 femtosecond). This experience served him well in later experiments using laser technology to "photograph" chemical reactions at femtosecond intervals. His findings are at the cusp of present research in physics and chemistry, and have had enormous influence in the scientific world. For his groundbreaking work in femtochemistry, he received the Nobel Prize in Chemistry in 1999. In the citation, the Nobel Prize Committee said: "This year's laureate in Chemistry is being rewarded for his pioneering investigation of fundamental chemical reactions, using ultra-short laser flashes, on the time scale on which the reactions actually occur. Professor Zewail's contributions have brought about a revolution in chemistry and adjacent sciences, since this type of investigation allows us to understand and predict important reactions."

Dr. Zewail's work has had a profound impact on chemistry and biology around the world. Experiments using the femtosecond techniques he pioneered are being used to examine processes on surfaces, to help improve catalysts; in liquids, where they help clarify the mechanics of reacting substances in solution; and in polymers, where they are being used to create new materials for use in electronics. In biology, the use of femtosecond lasers has demonstrated the ultra-fast beginnings of reactions, leading to new insights into the function of the retina and the uptake of oxygen into the blood. Many scientists predict that microbiology will be to the 21st century what physics was to the 20th, and if this is proven correct, Dr. Zewail's work will be viewed as a cornerstone of this new paradigm.

Dr. Zewail is also known for his lectures and his lucid explanations of complex physical and chemical processes. In this lecture, he begins with an overview of physics in the first decades of the 20th century, with particular attention paid to the nature of time. He then discusses the history of photographing moving objects, from the first crude attempts by Eadweard Muybridge through the more refined efforts of Étienne-Jules Marey, and then into the difficulties involved in making images at femtosecond intervals.

A major hurdle, and one which had led many sceptics to doubt that photography at femtosecond level was possible, was Heisenberg's uncertainty principle, which asserts that one cannot measure both the position and momentum of a particle at the same time. Dr. Zewail explains, in his characteristically straightforward manner, how this difficulty was overcome. Though one deals at molecular level with probabilities, these probabilities become wavefunctions which may be organized so the probability distribution is localized, acquiring the characteristics of a particle. It then becomes possible to make an image of the motion of the "particle".

Dr. Zewail then moves into a discussion of the uses of laser photography at femtosecond intervals to "see" chemical processes. At present, a new methodology, termed utrafast electron crystallography, is being used to examine the role of water in biological systems. The interaction of water molecules with proteins and DNA is of paramount importance in biology, and will have a profound impact on medicine. Dr. Zewail concludes by discussing some of the recent developments in femtotechnology, including the possible creation of extremely precise clocks and the possibility to induce nuclear fusion using techniques he developed.

The Ancient Library of Alexandria was at the forefront of the search for scientific knowledge. Here Eratosthenes, Hipparchus, and Euclid made lasting contributions to humanity. It is appropriate that a child of this region, who did his initial studies at the University of Alexandria, and whose work, like theirs, will continue to influence the global scientific community for a long time, should be our honored guest at the Bibliotheca Alexandrina for this outstanding lecture. Furthermore, we are delighted that he has been one of the founding members of the Board of Trustees of the Bibliotheca Alexandrina, guiding its initial steps.

Along with his scientific research, Dr. Zewail is noted for his humanitarian work. He makes it a point to communicate his knowledge through lectures, and has been instrumental in trying to bring a higher standard of education to the developing world. One of the stated aims of the Academia Bibliotheca Alexandrinae is to spread the values and culture of science in Egypt and the region, and it is a pleasure and an honor to welcome a true Alexandrian ambassador for science back to a city he once called home.

I. Serageldin

"...We need to understand the physics, the mathematics and the chemistry of biology, to make improvements in life sciences..." **Ahmed Zewail**

Professor Ahmed H. Zewail, born in Egypt in 1946, won the 1999 Nobel Prize in Chemistry for his groundbreaking work in "showing that it is possible with rapid laser techniques to see how atoms in a molecule move during a chemical reaction."

Professor Zewail is Linus Pauling Professor of Chemical Physics, and Professor of Physics at the California Institute of Technology (CalTech), and Director of the NSF Laboratory for Molecular Sciences. He is internationally recognized for his efforts in a field that he pioneered, femtochemistry. This technique uses ultra-fast lasers to probe chemical reactions as they actually occur in real time.

Because reactions can take place in a millionth of a billionth of a second, Zewail's research has, with state-of-the-art lasers, made it possible to observe and study this motion for the first time, thus allowing scientists to explore nature at its fundamental level.

Specifically, Zewail seeks to better understand the way that chemical bonds form and break. With the development of laser techniques, he and his team have been able to obtain greater insights about the exact nature of chemical bonds. The finding has had wide-ranging impact on chemistry and photobiology worldwide. Zewail's current research is devoted to dynamic chemistry and biology, with a focus on the physics of elementary processes in complex systems.

Professor Zewail's other honors include the Robert A. Welch Prize, the King Faisal Prize, and the Peter Debye Award. From Egypt, he received the Order of the Grand Collar of the Nile, the highest state honor; and postage stamps were issued to pay tribute to his contributions to science and humanity.

Professor Zewail's paper is entitled: "Time's Mysteries and Miracles: Consonance with Physical and Life Sciences"

Time's Mysteries and Miracles[*]:
Consonance with Physical and Life Sciences

Ahmed Zewail

Introduction

Ever since the dawn of history, humans have been the benefactors of time's miracles, but at the same time they have been baffled by time's mysteries. More than six millennia ago, the philosophy and measurement of time occupied the minds of scholars in the land of Bibliotheca Alexandrina, and, even today we struggle with the meaning of time. In this overview, I present some concepts and techniques developed in the science and technology of time, and an exposé of some of the mysteries and miracles that are in harmony with physical and life sciences.

Einstein spent a great deal of time thinking about time. In his theory of relativity, time is relative; its passage depends on how fast we travel relative to the speed at which light travels (300,000 km per second). In principle, time can be dilated and even stopped. Shakespeare knew this when he said "And time that takes survey of all the world must have a stop."

Perhaps the most puzzling issues, which have been with mankind for millennia, can be expressed in three questions: What is time? Why does it have a direction? How can it be resolved? The most complex question of all is the first one, because we really do not know what time is, and this leaves us with gray areas in the science and philosophy of time. One definition was given by C.J. Overbeck: "*Time is the great gift of nature which keeps everything from happening at once.*" Independent of its definition, we know that our perception of time depends on its duration, scale, and universality.

---

[*] Based on the Albert Einstein public lecture delivered in New Delhi, and adapted for the BioVision Nobel Laureates Day.

From the Microscopic to the Cosmic

All phenomena that we know of in our universe are defined by their time scales. Enduring or ephemeral in their character, these phenomena seem to follow an intriguing logarithmic scale of time that spans the very small (microscopic) world and the very big (cosmic) world. The human time-scale lies almost in between, a geometric average of the two extremes (Figure 2.1). The time of the big bang, the age of the universe, is about 12 billion years, or tens of 10^{+15} second (+15 on the log scale), recalling that one year is 32 million seconds. For the lightest atom, hydrogen, the time scale for the motion of an electron in its first orbit is about a tenth of a femtosecond, or a tenth of 10^{-15} second (-15 on the log scale). The average of the two limits is on the scale of seconds (zero on the log scale), the human heart beat–something to think about!

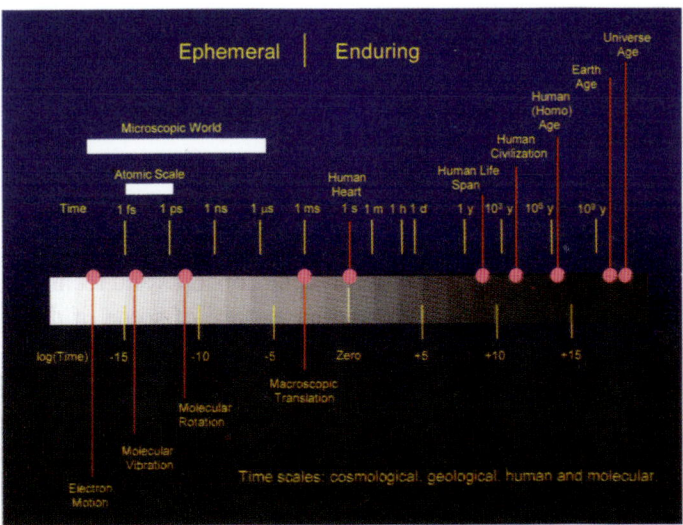

Figure 2.1. Time scales in cosmological, geological, human and molecular events.

On this log scale, we did not consider the ultimate–shortest–time of the universe, what is now known as Planck's time. In his attempt to give a universality to constants of nature, Planck in 1899 proposed that natural units of mass, length, time, and temperature can be constructed from the most fundamental constants: the gravitation constant G, the speed of light c, and the constant of action h (which now bears his name). By dimensional analysis, the shortest possible time becomes:

$$t(Planck) = (h\,G/c^5)^{1/2}$$

which is 10^{-43} seconds, and the corresponding length is 10^{-33} cm, obtained simply by multiplying by c. Even before 1900, the year quantum mechanics began to emerge, this unity in defining Planck's time is telling of "relationships" between quantum mechanics (h), gravity (G), and relativity (c). Implicit in this unification is the meaning of physical laws at scales below these values, and the nature spacetime with a universal speed of light – Einstein enters here!

Time, Light, and Relativity

Before Einstein, the great contribution by James Clerk Maxwell gave us a universal description of the nature of light. By a unification of electricity and magnetism, light, as a wave, propagates in space and time with electric and magnetic (electromagnetic) disturbances. This was a brilliant contribution expressed quantitatively in Maxwell's equations. Einstein in 1905 was concerned about two issues that relate to the nature of light–Is it really a wave? and, What happens to these waves if you can imagine running with them near the speed c? The first issue is not our concern here, but the second one is.

Something is special. Whichever direction a beam of light is coming from, independent of our own velocity for observation on Earth, we will always measure c for light. Einstein, in his special theory of relativity, gave the correct picture for adding velocities: For a motion of an object (say a moving ball) with velocity v in a reference frame (say a moving train) with a velocity u, an observer will see a motion not by the expected v + u velocity, but by v + u divided by the factor ($1+vu/c^2$); when the speeds v and u are the "normal" ones, that is, much less than c, then the total velocity is the expected (Newtonian) v + u. However, if instead of the ball we have light with speed c, then the total velocity becomes c + u divided by ($1 + u/c$) which is exactly c. The speed of light is the same in all reference frames, in all directions, for all observers, and every observer will experience the same natural laws.

The consequences of these findings for time, length, and mass require some philosophical interpretation. As the speed of light is approached, the length of a spaceship will shrink and approach zero in the direction of the motion. Similarly, moving objects become more massive and approach infinity when the object velocity becomes near the speed of light. For time, the mystery continues. Moving clocks must slow down and stop when the object velocity reaches the speed of light. In this "Dilation of Time," time becomes relative:

$$t(\text{moving}) = t(\text{stationary}) / (1-v^2/c^2)^{1/2}$$

where the velocity of the moving clock is v. From the expression, we note that the time of the moving clock gets longer (slowing down) as v increases, but we also note that if v is made to exceed c, we enter an imaginary world of time! Thus, within the framework of this theory, the speed of light is the ultimate speed in our world and universe.

In approaching these large scales of speed and mass, what happens to light? In his 1911-16 papers on the General Theory of Relativity, Einstein addressed the effect of gravity on light. Gravity is described as distortions in the four dimensions of space and time (three dimensions for space and one for time), and such distortions define Newton's "force" of gravity–spacetime is actually curved. Because of this curvature, a beam of light passing near the sun would bend in the gravity of a massive object. Experimentally, it was found by Arthur Eddington during the 1919 eclipse that indeed light was bent as it passed by the sun, as predicted by the theory. In 1922, Einstein received the 1921 Nobel Prize, not for his theory of relativity but for the photoelectric effect, a contribution that elucidated one of the two characteristics (duality) of light–a bundle of particles of quantized energy.

Symmetry of Time

Even if we consider the "normal world" when velocities, masses, lengths, and time are with no corrections–Newtonian Limits–and spacetime with no curvature, we still have problems with time, its direction and uncertainty. First let us consider the symmetry of time. Can time go forward and backward, or does it have a direction, an arrow?

In Newton's world, the motion of objects, like you and me, should follow "symmetry of time," that is, the equations describing motion on say the human scale, or that of the Earth around the sun, are time symmetric. There is no difference in the way they work if we make the direction of time go "forward" or "backward." Newtonian mechanics are deterministic and time symmetric. Because the force is related to the mass and the acceleration ($F=ma=m(d^2x/dt^2)$), the equation works equally well for positive and negative time. So, calculating the future of a physical system from its present situation is the same as calculating its past physical situation from its present one–weird and contrary to our common sense. What about microscopic systems, for example the world invisible to the eye–the atom?

For quantum systems, the equation of motion also has invariance under time reversal insofar as the positions of microscopic particles are concerned. This is true despite the deceptive appearance of a first derivative in the Schrödinger wave equation that would imply time reversal. If you can magnify a box containing a gas and see the atoms hitting each other individually you will conclude that there is no arrow of time for every pair of collisions. So, in Newton's mechanics and quantum mechanics, time flows in both directions, making the apparent confusion for the meaning of past, present, and future! In our life, we feel the passage of time and we also know that matter is made of atoms, so we have a dilemma.

Arrow of Time

Phenomena in our life follow an arrow of time. A cup of hot water with a piece of ice displays melting of the ice–the ice does not spontaneously reform again; heat always flows from a hotter object to a cooler one, and not the reverse. An egg breaks when it hits the floor, but it cannot be reformed from the floor. These and similar phenomena are described by the most powerful law, or what Arthur Eddington called the "supreme law of Nature"–the second law of thermodynamics. In one way it describes the arrow of time. In another way, it tells us about the content of information–there is a natural tendency for systems in change to become less ordered or more disordered. A measure of this change is called entropy which is defined as a negative measure of information. Entropy always increases (or at best does not change), order decreases, information decreases, and complexity decreases.

But this loss of information and increase in entropy is for the so-called closed systems (the ice and hot water form a closed system). In some cases, order is created of disorder, and it appears at first that this is in violation of the law of entropy. The tree is a good example–light from the sun, soil and water, and by photosynthesis we have an ordered tree. The Earth is not a closed system and is a part of the solar system–the local decrease in entropy for the tree is compensated for by the way solar (and other) energy change its entropy, and for the solar system on a whole, entropy is increasing according to the second law.

If entropy is always increasing in our universe and the arrow of time is well defined from past to future, why do individual particles, constituents of matter, follow trajectories that are symmetric in time? Put another way, why for a collection of particles each obeying time-reversal symmetry the ensemble as a whole defines an arrow of time? Imagine a box divided into two halves with a partition, one half contains a gas and the other is empty. If we remove the partition the gas will move and fill the whole box. Entropy increased and it appears that we can never reverse the process–we cannot make the gas go into one half and then reinstall the partition to acquire the originally ordered state. In the gas box each particle has a trajectory that follows Newton's mechanics. With time being symmetric, why then does the collection of these particles make the time unsymmetrical? This is a debatable subject and there are different views, one I find particularly interesting.

Time scales and recurrences in time

In the nineteenth century, Henri Poincaré considered this problem of a gas in a box, with all possible arrangements of the particles. He concluded that the system, *if we wait long enough,* will return back to the initial state. The time for this Poincaré recurrence is vastly different depending on the system under consideration. For the gas in the box, the recurrence time for reordering all particles is longer than the age of the universe, but for the vibrational motions of atoms and molecules it could be a millionth of a millionth of a second. This concept of time scale could explain the apparent behavior of systems, reversible or irreversible, depending on complexity and the number of possible arrangements or configurations.

This view is perhaps most clearly demonstrated on quantum systems with time scales short enough that we can experiment with them. If we take the same gas in the box and replace the hypothetical particles with shaped molecules we can perform an interesting experiment. To start with, we already know that there is no order in orientation of molecules and entropy is maximum. We now preferentially excite some of these molecules with their head and tails oriented roughly north and south of the box (we can do so in the laboratory with lasers). If the laser is ultrashort in duration (this too we can achieve in the laboratory) the induced ordered orientation of the molecules will ultimately be maximum at time zero and will decay with time. We call this process of degrading order dephasing, as the whole ensemble of millions of molecules prepared becomes out of step (phase) with each other. However, if we wait for some time, the molecules will acquire back the initial orientation giving rise to Poincaré's recurrences.

Such recurrences have been observed in our laboratory and on an ensemble of millions of molecules. Furthermore, these molecules are complex in their structure and internal motions and experts will tell you that these recurrences should not occur in such systems. But this is not true, as the energy levels are commensurate or nearly so even in complex systems. The recurrences are spaced long enough in time that depending on the time scale of observation the behavior of the system will appear differently. If the time scale of observation is too short, the system would appear irreversible in its decay behavior, but if we wait until recurrences occur we can then see the reversibility behavior.

Irreversibility becomes apparent if the system is not isolated. When the system interacts with a foreign perturber (such as collisions with other molecules–a heat bath) then such recurrences become weak and the system appears irreversibly disordered. Thus without designed methods for introducing order (coherence) to the system and/or without probes for observing its time evolution of disorder (dephasing) we may be misled about the nature of the dynamics. This is critical for defining the meaning and control of complexity and the time scale for reversible/irreversible behavior. We shall come back to this point when we consider measurement of time and matter's time scale.

The above consideration of microscopic/macroscopic behavior considers the origin of irreversible behavior in large ensembles as due to statistical "averaging." As such the law of entropy increase becomes a statistical law. To Ilya Prigogine, however, the second law of thermodynamics is a fundamental law describing irreversibility of nature–the gas in the box will never rearrange again and the ice in the hot cup will never reform, no matter how long we wait. We are now entering a risky area of interpretations and I prefer to stop here until we see further experimental proof! What about the behavior of individual atoms in molecules and their time scale? And, can we observe them moving with order in the ensemble?

At the Limit of Time–Democritus' Atom

The motion of atoms in molecules is fundamental to all dynamic changes of matter, whether the change is physical, chemical, or biological. But these atoms move with awesome rapidity and on ultrashort scales of time and length. On these scales, it is not clear that we can treat them as real, classical objects. Clearly, we must measure the passage of time for atoms on the time scale of the motion, and we must develop the concepts for understanding localization of atoms in space and time. Can this be achieved at the limit of time for quantum atomic motions? If we do, we will then observe Democritus' atom in motion and as a real object, making the transformation from the microscopic (wave function description) to the macroscopic (particle description) a reality in real time.

At Caltech, we have been interested in this endeavor of developing ultrafast laser light to freeze the motion of atoms, to make a motion-picture film of molecules with a frame resolution of a femtosecond. A femtosecond is a millionth of a billionth of a second. In one second, light travels 300,000 km (186,000 miles), almost the distance from the Earth to the Moon; in one femtosecond, light travels 300 nanometers, the dimension of a bacterium, or a small fraction of the thickness of a human hair. In principle, with femtosecond timing, the atom's motion becomes visible, but how can we advance stop-motion photography to reach the scale of the atom?

In the nineteenth century, the motion of animals was recorded for the first time using light shutters and flashes. In France, Étienne-Jules Marey, a professor at the Collège de France, was working (1894) on a solution to the problem of action photography using *chronophotography*, a regularly timed sequence of images. Marey's idea was to use a single camera and a rotating slotted-disk shutter, with exposures on a single film plate or strip that was similar to modern motion picture photography. Marey applied his chronophotographic apparatus in particular to humans and animals in motion, and to a subject that had puzzled people for many years: the righting of a cat as it falls so that it lands on its feet. How does the cat do it? Does its motion violate Newton's laws of mechanics? Does the cat have some special, magical physiology or a command of some weird new physics or what?

By "slicing time" and freezing the motion during the fall, in the transition state of the righting, Marey was able to answer the questions. First, the cat rotates the front of its body clockwise and the rear part counterclockwise, a motion that conserves energy and maintains the lack of spin, in accordance with Newton's laws. It then pulls in its legs, reverses the twist, and with a little extension of the legs, it is prepared for final landing. The cat instinctively knows how to move, and high divers, dancers, and some other athletes learn how to move in the absence of torque (the pushing force that gives you momentum in one direction or another). However, scientists needed photographic evidence of the individual stopped-action steps to understand the mystery. The answer to the puzzle was that the moving body was not rigid, and Newton's laws prevailed. At the time, these observations were thought-provoking and renowed scientists discussed in public their meaning and significance. J. Willard Gibbs gave a talk on December 4, 1894 before the Mathetical Club at Yale with the title "On motions by which falling animals may be able to fall on their feet." Marey's work and that of Eadweard Muybridge on the horse have changed the way we think of the behavior of animals (and humans) in motion.

For the world of atoms in molecules, if the above ideas of stop-motion photography can be carried over in a straightforward manner, then the requirements can be identified for experiments in femtochemistry—the field of studying molecular motions on the femtosecond time scale. The contrast in *length* and *time* scales for the motion of the cat and the atom is awesome (Figure 2.2). For a definition of 1 cm, a cat speeding at 2 m/s requires a time resolution of 0.005 second. But for a molecular structure in which atomic motions of a few angstroms (an angstrom, Å, is 10^{-8} cm) typically characterize chemical change, a detailed mapping of the motion will require a spatial resolution of less than 1 Å (about 0.1 Å). Therefore, the shutter time, or time resolution, required to observe with high definition atoms in motion at a speed of one kilometer per second (1000 m/s) is 0.1 Å divided by 1000 m/s, which equals 10^{-14} second or 10 femtoseconds—a million million times shorter than what was needed for Marey's (or Muybridge's) stop-motion photography. Although this was a central idea in the development of femtochemistry, we had to overcome a major dogma regarding the uncertainty principle!

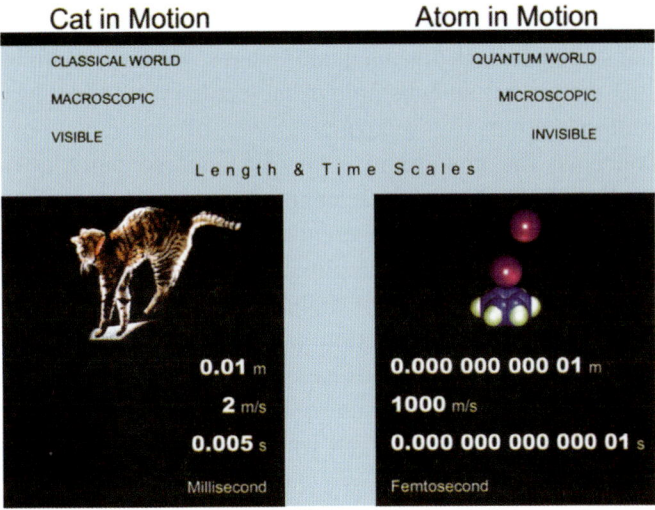

Figure 2.2. Length and time scales of atoms and cats.

Solving the Riddle of Uncertainty–Physics

For the atom such minute time and distance scales mean that molecular-scale phenomena should be governed by the principles, or language, of quantum mechanics, which are quite different from the familiar laws of Newton's mechanics that were used in the description of the motion of the cat and horse. Werner Heisenberg in the 1920s discovered that for quantum systems we are not allowed to make a precise measurement of both the position (x) and the momentum (p) of a particle at the same time. This tells us that we are losing knowledge – we do not know exactly where it is and where it is going (future), simultaneously, that is, the more accurately we determine one of these conjugates the more information we lose on the other. There is intrinsic uncertainty! Similarly, if we can measure the energy (E) of a system very precisely we cannot obtain the same precision for time (t) simultaneously. There is uncertainty in the measurement of time depending on how accurate the energy is, and the consequences are important for all sciences on the ultrashort time scale.

These considerations of uncertainties led initially to the belief that the femtosecond time resolution would not be useful. Moreover, predictions suggested that localization of atoms in space–wave packets–would not be possible to sustain for a long time, even on the femtosecond scale. Finally, there is a fundamental difference in the analogy between femtosecond stop-motion action of atoms and the millisecond photography of a cat or horse-in femtochemistry experiments one probes typically millions to trillions of molecules, and/or repeats the experiment many times to provide a signal strong enough for adequate images. Unlike experiments on one cat or one horse, the picture for an ensemble of molecules would be blurred.

We accommodate this by recognizing two of the most powerful and yet indigestible concepts: the uncertainty principle and the particle-wave duality of matter (de Broglie, 1924). The complementary aspect of these two descriptions is interwoven with the concept of coherence. Two or more waves can produce interference patterns when their amplitudes add up coherently. For matter, superpositions analogous to those of light waves can be formed from matter wavefunctions. The Schrödinger equation yields wavefunctions together with their probability distributions, which are diffuse over position space. But if these waves are added up coherently with well-defined phases, the probability distribution becomes localized in space. The resultant wave packet and its associated de Broglie wavelength has the essential character of a classical particle: a trajectory in space and time with a well-defined (group) velocity and position–a moving classical marble but at atomic scale.

To see motion in real systems, localized wave packets must form in every molecule, and there must also be a limited spread in position among the wave packets formed in the millions of molecules studied. This is achieved by the well-defined initial equilibrium configuration of the molecules before excitation and by the "instantaneous" femtosecond launching of the packet. The spatial confinement of the initial ground state, typically 0.05 Å, ensures that all molecules, each with its own coherence, begin their motion in a bond-distance range much smaller than that of the actual motion, typically 5^{-10} Å. The femtosecond launching ensures that this narrow range of bond distance is maintained throughout preparation. With coherent and synchronous preparation, the motion of the ensemble becomes that of a single-molecule trajectory.

In 1987, we reached our goal of observing, for the first time, Democritus' atom—theorized by the Greek philosopher some 2500 years ago–in motion, and we could describe it on the femtosecond time scale as a classical object like the cat and horse (Figure 2.3). In reaching the femtosecond domain of the atom, with a scale of a millionth of a billionth of a second, the time resolution of today compared to that of a century ago, with a scale of a thousandth of a second, is like one day compared to the age of the universe.

Eugene Wigner and Edward Teller debated the uncertainty paradox for picosecond time-resolution in a lively exchange at the Welch Conference in 1972. But, because of coherence, the uncertainty paradox is not a paradox even for femtoscience, and certainly not for the dynamics of physical, chemical, and biological changes. Charles Townes encountered objections in the realization of the maser because of concern about the uncertainty principle, but coherence was again the key to success. As we cross the femtosecond barrier into the attosecond regime for studies of electron dynamics. we must recall this vital role of coherence. Otherwise the spectre of quantum uncertainty might veil the path to new discoveries.

In retrospect, this vital role of coherence in the uncertainty paradox and the fog that surrounded its utility should have been clear (Figure 2.3 and bibliography). We and others have considered in detail the theoretical quantum calculations of molecular systems and indeed confirmed the localized motions of atoms. But, the physical origin of the behavior is simple to understand. Considering the uncertainty in the position to be Δx, and similarly for the other variables, the two uncertainty relations,

$$\Delta x \, \Delta p \geq \hbar /2$$
$$\Delta t \, \Delta E \geq \hbar /2$$

show that the only way to localize atoms (small Δx) is by shortening time (Δt). Moreover, when Δt is on the femtosecond time scale, even a discrete quantum system, if excited coherently, becomes effectively a continuum or quasi-continuum of energy states, which represents a transition to the classical world.

Given that we can localize a system to an initial distance of Δx_0 at time zero, why does the system remain coherent and behave as a classical object? And, does the time for the loss of coherence depend on the size of the object? Because the value of $\hbar$ is very small, this time depends crucially on the size. To see this clearly, we must recall that the uncertainty relation relates the uncertainty in position (Δx) to the uncertainty in momentum (Δp); but it is the velocity, and not momentum per se, which tells us the future position. Since $\Delta p = m \, \Delta v$, it follows, from the uncertainty relation, that $\Delta v = \hbar /(2m \, \Delta x_0)$–the larger the size (the larger the mass m and also the larger the scale of precision in position Δx_o) the smaller the uncertainty in velocity (Δv) and the better we are in predicting the future. Now it is straightforward to calculate the "time of uncertainty" which tells us how long it will be before the uncertainty in velocity will contribute as much to our lack of knowledge of where the object is as that which came from the original position uncertainty (Δx_o):

$$t \text{ (uncertainty)} = \Delta x_o / \Delta v = 2m \, \Delta x_o^2 / \hbar$$

Beyond this time scale, the uncertainty, due to our lack of knowledge of velocity, makes us less certain of the future and the description of the object becomes quantum, not a classical one. This simplified equation can be obtained from a more rigorous treatment of wave packet motion, and elsewhere we did so.

The size of $\hbar$, 1×10^{-27} erg-sec, means that the fuzziness required by the uncertainty principle is imperceptible on the normal scales of size and momentum, but becomes important at atomic scales. For example, if the position of a stationary 200-g apple is initially determined to a small fraction of a wavelength of light, say $\Delta x_o = 10$ nm, the apple's position uncertainty will spread by about 40% only after 4×10^{17} s, or 12 billion years, that is, the age of the universe! On the other hand, an electron with a mass 29 orders of magnitude smaller would spread by 40% from an initial 1-Å localization after only 0.2 femtosecond.

From atom to man, the time and length of uncertainty determine the classical-quantum description of motion (Figure 2.4). The time scale for future uncertainty runs from femtoseconds for the hydrogen atom, to 300 years for biological cells, and to more than the age of the universe for humans–we have 300 years (or more) to behave in a deterministic classical world, so biotechnologists can be sure to improve the human life expectancy by at least three times from the current one without the need of new mechanics!

Figure 2.3. Uncertanities and unification through coherence.

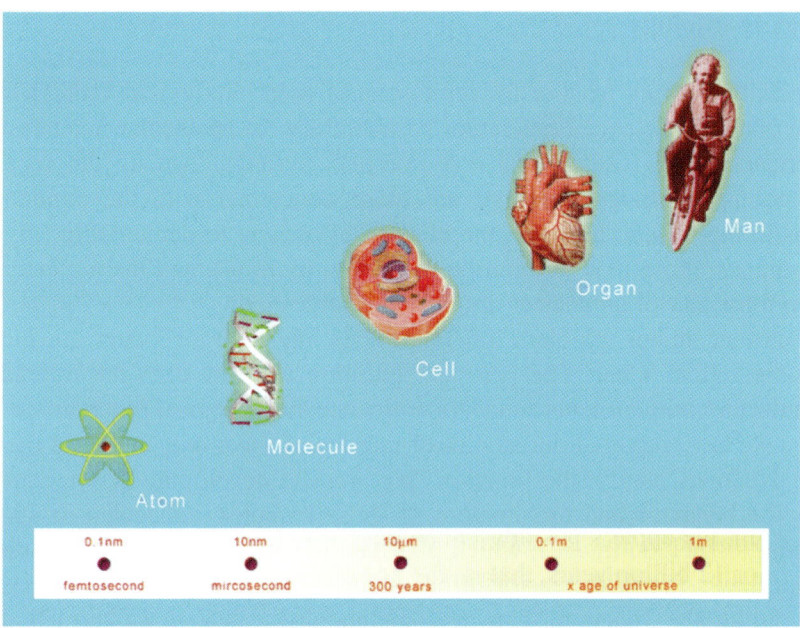

Figure 2.4. From Atom to Man- length scale and time of uncertainty.

The Molecular World–Chemistry

Conceptually, our work in the late 1970s on coherence phenomena and in the mid 1980s closing in to resolve reaction dynamics in real time provided the foundation for thinking about the issues raised above. It became clear that molecules can be made to vibrate coherently and ensembles of molecules can be made to behave in unison. Experimentally, we needed a whole new apparatus, a whole new "camera" with unprecedented time resolution. We needed to interface femtosecond lasers and molecular-beam technology, which required not only a new initiative but also a major effort at Caltech. In a relatively short time, femtochemistry research became active in many laboratories around the world.

The breadth of applications emerging spans the very small to very complex molecular assemblies, and all phases of matter. An example that demonstrates the unity of concepts from small to large molecular systems came from a paradigmatic study made at Caltech on a sibling of table salt (two atoms) and another at Berkeley on the protein molecule of vision (hundreds of atoms). In both, the primary step involves femtosecond motion of the atoms, and we now understand better the remarkably coherent and highly efficient first step of vision at the atomic level.

Complexity–Biology

An especially exciting frontier for femtoscience is in biology. At Caltech we now have the National Science Foundation's Laboratory for Molecular Sciences (LMS) for interdisciplinary research on very complex systems. Among the recent new studies published in femtobiology are those concerned with the conduction of electrons in the genetic material, the binding of oxygen to hemoglobin (myglobin) and its mimics, molecular recognition of protein by drugs, and the molecular basis for the cytotoxicity of anticancer drugs, and of digestion.

A current major problem of interest is the role of water in biological systems–biological water. The pertinent question is: How does the interaction of water molecules with proteins and DNA influence the biological function? In a series of papers we have reported our studies of interfacial water dynamics and the unique role the dynamics play in the function. We are also developing new techniques to observe the behavior and architecture of these complex molecules–in space and time–using diffraction images, which give the 3-D location of all the atoms, all at once. But now a fourth dimension–time–is introduced to see how complex systems behave during the function. The new methodology, which we termed ultrafast electron crystallography (and microscopy), is now established with many applications (see bibliography). The impact on biology and medicine is clear.

Life is a manifestation of complexity in which atoms of the microscopic world combine in different ways to form functional systems with enormous diversity and unique information. And that is what makes the human "intermediate scale" (Figure 2.1) special–on one hand simple in function and on the other hand rich in complexity. Deciphering this complexity and reducing its meaning to the atomic motions involved is one of the most fundamental problems of this century.

Technology of Femtoscience

As for technology developments–femtotechnology–there are exciting new developments in microelectronics (femtomachining), femtodentistry, and femtoimaging (microscopy) of cells and tumors, not to mention possible new developments with intensities reaching that of the sun (in femtoseconds!) and duration going beyond the femtosecond (attosecond), and the interface with nanoscience and technology–marrying scales of time and length. The ability to count optical oscillations of more than 10^{15} cycles per second will lead to the construction of all-optical atomic clocks, which are expected to outperform today's state-of-the-art cesium clocks, with a new precision limit in metrology. There is also the potential for using powers reaching 10^{20} watts/cm^2 to induce nuclear fusion in clusters of atoms through Coulomb explosion. There is also the possibility for controlling matter on the femtosecond time scale–one day we may direct chemical reactions into specific or new products.

Epilogue

I wish to conclude by conjecturing on some future mysteries and miracles of time. In the physical sciences, one advance that surely will allow us to reach the electron domain involves measurements on the sub-femtosecond time scale. Now the average energy is nearing the x-ray region, much above chemical and biological energies, and the pulse width is larger than chemical binding energies. Nonetheless, such advances will make it possible to study electron dynamics in many domains of physics and related areas.

In the life sciences, the advent of diffraction and microscopy techniques with atomic-scale spatial and temporal resolution will undoubtedly lead to a revolution in structural dynamics of biomolecules, building real bridges between structures, dynamics and functions (see bibliography).

In cosmology, Planck's scale of time, the nature of spacetime, and the arrow of time are subjects that will remain in need of further discovery and search for meaning.

From the very small (atom), to the very complex (life), to the very big (universe), despite some mysteries, new frontiers will be reached with time defining a fundamental dimension. Perhaps the biggest of all challenges is reversal of time. Ever since H. G. Well's novel "The Time Machine," the human imagination has considered the possibility of reversing the arrow of time, going back in time. In theory we could, but the paradoxes are many. A time traveler may go back in time and alter circumstances leading to his own existence or lack thereof. Two-way time travel is indeed weird, and may force an entry to the world of weird physics! So despite its miracles and the impact on our life, we still struggle with the meaning of time.

Bibliography

Albert, David. 2000. Time and Chance. Harvard University Press, Cambridge.

Barrow, John D. 2002. The Constants of Nature. Pantheon Books, New York.

Gribbin, John. 1998. In Search of the Big Bang (new edition), Penguin Books, New York.

Hall, Nina. 2002. Chemistry and Biology in the New Age. Chem. Commun. (Feature Article, Ahmed Zewail), October 7 issue, 2185.

Hazen, Robert M., and James Trefil. 1996. The Physical Sciences. John Wiley & Sons, New York.

Ihee, H., V. Lobastov, U. Gomez, B. Goodson, R. Srinivasan, C.Y. Ruan, and A.H. Zewail. 2001. Direct Imaging of Transient Molecular Structures with Ultrafast Diffraction. Science, 291, 385.

March, Robert. 1978. Physics for Poets, Contemporary Books, Chicago.

Ruan, C.Y., V. Lobastov, F. Vigliotti, S. Chen, and A.H. Zewail. 2004. Ultrafast Electron Crystallography of Interfacial Water. Science, 304, 80.

Zewail, A.H. 2004. Femtochemistry–Atomic-Scale Dynamics of the Chemical Bond Using Ultrafast Lasers. Les Prix Nobel, The Nobel Prizes, Almquist & Wiksell International, Stockholm.

Zewail, A.H. 2001. The Uncertainty Paradox–The fog that was not. Nature, London, 412, 279.

Zewail, A.H. 2002. Voyage Through Time–Walks of Life to the Nobel Prize. American University Press, Cairo (also in 12 editions and languages).

Zewail, A.H. 2005. Diffraction, Crystallography, and Microscopy Beyond 3D–Structural Dynamics in Space and Time. Philosophical Transactions, Royal Society, 364, 315.

Science and Technology
in the
Twenty-First Century

Ahmed Zewail
Nobel Laureate

AKADEMI
SAINS
MALAYSIA

INTRODUCTION

Assalamualaikum w.r.t.,
Ladies and Gentlemen

The title of my lecture implies that I will make predictions about science and technology (S & T) in the twenty-first century, but I know better – one should be cautious about making predictions as many in the past have proven to be off the mark. Take, for example, the prediction made in 1943 about the future of the computer by Thomas J. Watson, founder of IBM: "I think there is a world market for about five computers." In retrospect, this is a remarkable statement given that today we have billions of computers!

It is in the nature of science that we scientists search for the truth in the unknown, which is so vast and complex that our predictions will always be constrained by our ignorance of the future. The renowned historian of science Karl Popper described the state of knowledge this way: "Our knowledge can only be finite, while our ignorance must necessarily be infinite." The famous actor Woody Allen cast the same sentiment in a more humorous light: "Can we actually 'know' the universe? My God, it's hard enough finding your way around in Chinatown." Indeed, it is hard to predict the future, but it is hoped that with insight we can ask the right questions, gain new knowledge, and develop new technologies. In this process of scientific research it is essential that we understand the culture of science and its foundations – inherent continuity, uncertainty, and unmanageability.

SCIENCE AND ITS METHODOLOGY

Since the beginning of human civilization, science and technology has progressed in a continuous process. Fire must have been an exciting new technology for the first humans and to this day we are continuing research to fully answer the question, what is fire? But the search for new knowledge is based on *rational thinking*, which is fundamental for progress and for making new discoveries. I doubt whether there was (or now is) a civilization that reached a high level of achievement without simultaneously nurturing S & T and employing the rational thinking characteristic of the culture of science. After all, we are *Homo sapiens*, the species characterized by an enlarged brain capacity.

Science is an education process that allows the educated and creative minds to question, experiment or observe in an attempt to find answers, and then try to identify a set of unifying principles, concepts, and laws that embraces all phenomena of nature. The aim is to better understand our universe and gain new knowledge that will enlighten humanity by unveiling mysteries of how nature works. In the process we may make new discoveries and inventions that change the way we think and/or create new technologies that transform our society.

The sharp division of science into pure and applied branches is not natural. Some managers of science believe in this division and wish to emphasize only "what is relevant" for the prosperity of the society. But that is not the way science works, as scientists themselves in their quest for new knowledge do not know what is relevant. And if they knew ahead of time it would not be new knowledge. Scientific research is not manageable in the usual sense of the word.

THE TRIAD – THE CLONING CASE

Most S & T advances are governed by a structure of a connected triad – basic research, technology development, and the involvement of society. In this cycle, both pure and applied science become an integral part of a successful endeavor. Take the case of cloning, a current subject of significant relevance to the definition of *species*. Cloning began as a laboratory experiment on the genetic material of our cells, DNA. But these experimental achievements would not have been possible without scientific research and advances over a half century of development in many areas: the discovery of the genetic code, molecular structure of DNA, recombinant DNA, and other related studies. Moreover, a century of development of new techniques and tools provided the means: x-ray crystallography, polymerase chain reaction (PCR), genetic engineering, as well as computer data processing.

Then came the first successful cloning of a higher organism – Dolly, the sheep – an act that transformed laboratory research into an enterprise with possible benefits to human medicine. But the third element of the triad, the society, must now be involved to develop a full perspective of the moral, ethical, and religious implications of cloning. With rational thinking, the benefits of cloning and other developments, such as stem cell research, to society will undoubtedly feed back to the support of basic research, and the cycle of the triad continues.

REVOLUTIONARY PARADIGMS

Cycles of this type ultimately lead to the development of new concepts and new tools or techniques. Some historians of science make a division between the two. The late influential historian and philosopher of science, Thomas Kuhn, favored *concept-driven* research as a paradigm over *tool-driven* research. Although I am not a historian, I find, from my own experience as a scientist, this distinction is not as sharp as Kuhn would have it. In fact, new tools and techniques drive scientific research as much as new concepts, and both are part of the progress of science.

Gravity, relativity, and quantum mechanics are concepts that surely have changed the way we think. With these concepts we can describe phenomena of nature and understand their fundamentals – objects attract each other depending on their masses and the distance between them (gravity); objects that are massive or move at very high speed do not follow Newton's classical laws of mechanics (relativity), and objects of microscopic size also do not follow the laws of Newton (quantum mechanics). But these concepts and rules need to be tested experimentally and the phenomena have to be observed and studied. Moreover, in most cases the concepts and rules develop as a result of experimentation and observations that are made possible with new tools and techniques. In fact, one may argue that techniques for experimentation and observation are prerequisite for the development of concepts. Of course, these techniques must be used creatively in order to develop new paradigms and concepts.

CAMERA OBSCURA, MICROSCOPE, AND TELESCOPE

The microscope, the telescope and the simple *camera obscura* were revolutionary not only because they made it possible to observe the very small and the very far, and to record a picture of what is viewed, but also because they changed the way we think of biology, astronomy, and vision. The conceptual pioneer of *camera obscura* was ibn al-Haytham (known in the West as Alhazen; ca. 965–1038), a Muslim polymath acknowledged to be the greatest scientist of the European Middle Ages. The principle of *camera obscura* (among other observations) explained at the time the nature of light – it travels in straight lines, it has high speed, and it reflects from bodies (and refracts in media) into the 'human camera', the eye. Alhazen's work in optics caused a paradigm shift about vision – a reflection of light from objects to the eyes instead of radiating light from them – and established the foundation for the new technologies of still and motion picture photography.

The light microscope was developed in the middle of the seventeenth century. In Holland, Antoni van Leeuwenhoek (1632–1723) and in England Robert Hooke (1635–1703) made astounding discoveries, including observation of tiny moving creatures in droplets of water, sperms, and the cellular structure of slices of cork. Hooke coined the word *cell*, and his greatest work, *Micrographia* (1665), defined microscopy as a scientific discipline. As a result of these advances, the history of biology has shifted from an emphasis on classifying living organisms and plants to studying the living cells – the exploitation of new tools produced cell biology as a new branch of science. Molecular biology and genetics are the most recent frontiers reached with the aid of other developments based on the use of electromagnetic waves, those of x-ray diffraction by DNA and protein crystals and nuclear magnetic resonance (NMR) of macromolecules. Through scientific experimentation the microscope has magnified the world of the very small – microns in size – so it is visible to our eyes. As a result human medicine has changed forever.

The telescope was invented before the microscope, in the early part of the seventeenth century; some believe that the first optical assembly of this nature was made in 1550. Hans Lippershey, an eyeglass maker based in Holland had developed telescopes (1608-09) with a magnifying power of about three times. These and later telescopes were made from a combination of concave and convex lenses and the effect produced was understood from studies of the refraction and reflection of light (e.g., those by Alhazen and later by the Dutch scientist Willebrord Snell). Galileo Galilei (1564–1642) was the first to make use of the telescope to visually approach the very far, first of ships and then the heavens. Through scientific observations (1610) of Jupiter's moons, which revolve around a planet other than Earth, Galileo refuted the long-held dogma of geocentrism, proving that the stationary Earth is not at the center of the universe with the planets and sun revolving around it. The geocentric model – Ptolemaic astronomy at the heart of Aristotelian world view – was to be replaced by the heliocentric model of Copernicus (1473–1543), in which the sun is the fixed center of the universe and the planets, including Earth, are in circular orbits around the sun; Kepler (1571–1630) refined the Copernican model by showing that astronomical bodies follow elliptical orbits in their motion. Without these techniques and concepts we could not launch a spaceship or a satellite or hope to understand our universe.

FEMTOSCOPE AND MATTER'S ULTRAFAST UNIVERSE

Resolution of time is not a property of the microscope or the telescope, and to study ephemeral phenomena of matter we needed new developments. At Caltech, the integration of new techniques and concepts led to the development of the "femtoscope", which brings to vision the motion of atoms in real time. In so doing we are able to uncover the underlying fundamentals of the dynamics in simple and complex systems, and examine the forces responsible for the diverse molecular functions in diseases such as cancer, in the sense of taste, and in the elementary mechanism for food digestion.

In the microscopic world – the quantum world – the motion of atoms can be observed only with an 'artificial eye' having a speed one million million times faster than our naked eye, which responds in a fraction of a second. Microscopic motion is ephemeral and ultrashort in duration, and we need a telescope that not only brings their very far world up close for observation, but also freezes them in time so we can take snapshots. Ultrafast laser optics is the essential element of this "femtosecond telescope", the femtoscope. A femtosecond is 10^{-15} second or 0.000 000 000 000 001 second. In one second, light travels about 300,000 kilometers, almost from Earth to the moon; in one femtosecond, light travels 300 nanometers (0.000 000 3 meter), the dimension of a bacterium, or a small fraction of the thickness of a human hair.

In principle, with femtosecond timing, the atom's motion becomes visible, but how can we advance stop-motion photography to reach the scale of the atom? In the nineteenth century, the motion of animals was recorded for the first time using light shutters and flashes. In Paris, Étienne-Jules Marey, a professor at the Collège de France, was working ·(1894) on a solution to the problem of action photography: chronophotography, a reference to the regular timing of a sequence of images. Marey's idea was to use a single camera and a rotating slotted-disk shutter, with exposures on a single film plate or strip that was similar to modern motion picture photography. Marey was interested in investigations of humans and animals in motion, including a subject that had puzzled people for many years, namely the righting of a cat as it falls so that it lands on its feet. Marey's work and that of Eadweard Muybridge on the horse have changed the way we think of the behavior of animals (and humans) in motion.

For the world of atoms in molecules, if the above ideas of stop-motion photography can be carried over in a straightforward manner, then the requirements can be identified for experiments in femtochemistry – the field of studying molecular motions on the femtosecond time scale. For a molecular structure in which atomic motions of a few angstroms (an angstrom, Å, is 10^{-8} cm) typically characterize chemical reactions, a detailed mapping of the reaction process will require a spatial resolution of less than 1 Å (about 0.1 Å). Therefore, the shutter time, or time resolution, required to observe with high definition the molecular transformations in which atoms move at speeds of the order of one kilometer per second (1000 m/s) is 0.1 Å divided by 1000 m/s, which equals 10^{-14} second or 10 femtoseconds – a million million times shorter than what was needed for Marey's (or Muybridge's) stop-motion photography.

However. such minute times and distances mean that molecular-scale phenomena should be governed by the principles, or language, of quantum mechanics, which are quite different from the familiar laws of Newton's mechanics that were used in the description of the motion of the cat and horse. Conceptually and experimentally we addressed these issues, and in 1987 we reached our goal of observing in motion, for the first time, Democritus' atom – theorized by the Greek philosopher some 2500 years ago – and we could describe it on the femtosecond scale as a classical object like the cat and horse. The similarity between atomic motions and planetary classical motions brings about an analogy between the femtoscope and the telescope. In reaching the femtosecond domain of the atom, with a scale of a millionth of a billionth of a second, the time resolution of today compared to that of a century ago, with a scale of a thousandth of a second, is like one day compared to the age of the universe.

With the femtoscope, the breadth of applications emerging from all over the world spans very small to very complex molecular assemblies and all phases of matter. An example that demonstrates the unity of concepts from small to large molecular systems came from a paradigmatic study case made at Caltech on a sibling of the table salt (two atoms) and at Berkeley on the protein molecule of vision (hundreds of atoms). In both, the primary step involves femtosecond motion of the atoms, and we now understand better the remarkable coherent and highly efficient first step of vision at the atomic level.

An especially exciting frontier for femtoscience is in biology. At Caltech we now have the National Science Foundation's Laboratory for Molecular Sciences (LMS) for interdisciplinary research on very complex systems. Among the recent new studies published are those concerned with the conduction of electrons in the genetic material, the binding of oxygen to models of hemoglobin, molecular recognition of protein by drugs, and the molecular basis for the cytotoxicity of anticancer drugs and of digestion. We are also developing new techniques to observe the behavior and architecture of these complex molecules – in space and time – using diffraction images, which gives the 3-D location of all the atoms, all at once! The impact on biology and medicine is clear.

As for technology developments – femtotechnology – there are exciting new developments in microelectronics (femtomachining), femtodentistry, and femtoimaging of cells and tumors, not to mention possible new developments with intensities reaching that of the sun (in femtoseconds!) and duration going beyond the femtosecond (attosecond), and the interface with nanoscience and technology – marrying time and length scales. The ability to count optical oscillations of more than 10^{15} cycles per second will lead to the construction of all-optical atomic clocks, which are expected to outperform today's state-of-the-art cesium clocks, with a new precision limit in metrology. There is also the potential for using powers reaching 10^{20} watts/cm^2 to induce nuclear fusion in clusters of atoms through Coulomb explosion. And, the possibility of controlling matter on the femtosecond time scale – one day we may direct chemical reactions into specific or new products. None of these developments, applications, and technologies were strategically planned and managed.

FUTURE FRONTIERS

With these reflections on the culture and progress of science, what do I expect for S & T in the twenty-first century? I see three major frontiers that through the triad of basic research, technology development, and the involvement of society will be rich in new paradigms and have direct impact on human life.

Our matter – the scale of the very small. We are on our way to being able to manipulate matter at its smallest, most fundamental limits, both in time, on the femtosecond scale, and in length, on the nanoscale. Just think about these new scales of time and space in the world of the very small. If your heart beats once a second, now we can see the beats of atoms in a femtosecond, in a millionth of a billionth of a second – a femtosecond is to a minute as a minute is to the age of the universe. Similarly, we can study matter on the nanometer scale and resolve the atoms in their structures – the size of the atom to the size of the earth is like the size of the earth to the whole universe. The opportunity is huge for acquiring new knowledge and for creating new forms of "our matter." The manipulation of matter to produce new sources of energy (photovoltaic/ photosynthetic, etc.) should become a major undertaking. The interface of matter's micro- and nanonetworks, designed to produce artificial intelligence, to our life organs, such as the brain, will be another frontier that could alter the boundaries and meaning of species.

Our universe – the scale of the very big. In this century, we may have colonies on the moon, and we may have our second homes on other planets and maybe even in other solar systems. Just think of the scales of the world of the very big. Our universe is about 12 billion years old, and at the speed of light (300,000 km/s), our universe's limit of distance is 100 billion trillion kilometers – certainly enough space for the six billion people on earth today, even multiplying by ten or by one million for the future! The opportunities involving outer space and information technology are unlimited. On our planet, through 'virtual walls', which in principle will provide any information one needs, education and intelligence in all societies will be redefined.

Our life – the scale in between. In the first year of this century, the sequencing of the human genome was completed. We now have the genetic map that describes every human on planet Earth. Just think, three billion letters have been deciphered and read into our book of life. The history of biology has changed from the classification of living organisms (Darwin's theory), to the world of cells (Leeuwenhoek–Hooke microscope), to now the molecular world (Watson and Crick's DNA) with revolutionary ideas in genetic engineering. Soon we might see a nanoscale motor entering the cell to do work. Medicine and human health will certainly enter a new age.

CAUTIONARY REMARKS

The scope of opportunities is wide ranging and the promise in these new and other areas of science for new discoveries and new technological developments are becoming a reality. But I have a few cautionary general remarks to make.

First, in the new mode of research in the twenty-first century, we should not become professional technical experts in narrow areas of specialization. In this regard, multidisciplinary science may help, but should not be at the expense of depth and scholarship.

Second, the mix of science and business by scientists is a concern that in this century may, in my opinion, have a detrimental impact on the culture of science and commitment to scholarship. Academia should remain the place for free exchange, motivated primarily by the search for new knowledge and education of students.

Third, the support of research should be granted for the best ideas and the best people, not for strategic- or mission-oriented, and managed, research. James B. Conant, the renowned educator and scientist and former President of Harvard University, in a 1945 letter to The New York Times commented: "There is only one proven method of assisting the advancement of pure science – that of picking men [and women] of genius, backing them heavily, and leaving them to direct themselves." Naturally, there should be other places for carrying out research and development for specific national missions, but institutes of knowledge should not be managed and missioned. Investment in science and education is the best thing a nation can do for posterity.

THE WORLD OF THE HAVE-NOTS

For the developing nations, there are barriers for progress, but I see no way out of investing in science education and science development. A look at global S & T is illuminating. The total number of scientific papers published worldwide over the past five years is 3.5 million. The European Union's share is 37%; the United States, 34%; Asia Pacific nations, 21%. The United States contributes 30% to 40% to the world economy and the strong correlation between the advanced state of S & T and the advanced state of the nation is clear. Asian Pacific countries are showing exponential growth in S & T papers and this too explains their increased position in the world economy. Other correlations of gross domestic product (GDP), health status, life expectancy, and illiteracy show the critical role of science education and scientific research in the global positioning of nations.

The lack of a solid science and technology base is not always a result of poor capital or human resources. It sometimes stems from a lack of appreciation of the critical role of S & T in development, an incoherent methodology for establishing such a base, and an absence of a coherent policy addressing national needs, and human and capital resources. Some countries consider scientific progress to be a luxury, as measured against other demanding concerns. Others believe that the base can be built by buying technology from developed countries. These beliefs translate into poor or, at most, modest advances that are based on the efforts of individuals, not institutional teamwork.

These issues point to three essential ingredients for progress. First is the building of human resources by eliminating illiteracy, ensuring active participation of women in society, and reforming education. Second is to rethink national constitutions, allowing for freedom of thought, minimizing bureaucracy, developing a merit system, and creating a credible – and enforceable – legal code. Third is the building of a science base.

The foundations of a science base are investment in special education for the gifted, the establishment of centers of excellence, and the chance to apply knowledge in the industrial and economic markets of the country and, eventually, the world. This must go hand-in-hand with a plan for general education at state schools and universities. With such a vision, a scientific culture will emerge that enhances a country's ability to follow and discuss complex problems, rationally and collectively. Scientific thinking becomes essential to the fabric of the society.

Developing countries need centers of excellence, not only for research and development, but also for training experts in advancing technologies and so reducing the brain drain experienced by many such countries. It is important that these centers are not just exercises in public relations: They should be limited to a few areas of research in order to build confidence and recognition. In the coming fifty years, knowledge-based and skill-based societies will have the lion's share of the world market and high status. Without S & T how can the have-nots participate in current world issues such as stem-cell research, cloning, human genome sequencing, artificial intelligence, manipulation of matter, molecular medicine, and cosmology? Without S & T how can they actively contribute to the world market in technologies such as microelectronics, information and communication, new materials, and the revolutionary biotechnologies?

The challenges require a new system of education and a new outlook on technologies. Technologies fall into three categories, those that are 'simple' but relevant to services, solving domestic problems of everyday life, from traffic lights to desalination of water; those that are 'innovative', which make participation in the world market possible, such as microelectronics; and those that are 'frontier', which are concerned with research into the unknown, representing an investment in the future. To be effective, a new system of education and research and development in the first two categories are required and, at the least, there must be serious engagement with the issues of the third, frontier category – where the world is going to be. Developing countries must address these issues with a new action plan and serious commitment.

But it is also the responsibility of developed countries to focus aid programs. Usually an aid package is distributed across many projects, with a lack of follow-up that leads to a diffusion of resources and in some cases corruption, so the aid does not result in significant successes. Real focus can be achieved by establishing what I call "partnership-guided aid", with a major portion of the aid being directed toward excellence using criteria established in developed countries.

There must also be a minimization of politics in aid. The use of an aid program to help specific regimes or groups is a big mistake, as history has shown that it is in the best interests of the developed world to help the the people of developing countries. Aid programs should be visionary in their mission and supportive of investment in future developments. Developed nations either can give money as charity write-offs or they can become partners, providing expertise and a follow-up plan.

EPILOGUE

Let me end on a hopeful note. If humanity's quest is progress and prosperity, we will need to weave the rational scientific approach, which is basic to our definition as *Homo sapiens*, into the fabric of our civilizations; if we do so, I can predict a bright future. Surely, science and technology will then become the real spaceship for the successful voyage of posterity.

Nobel Laureate Dr Ahmed Zewail during his courtesy visit to
Rt. Hon. Dato' Seri Dr Mahathir Mohamad, Prime Minister of Malaysia
at Putrajaya on 15 October 2002

Nobel Laureate Dr Ahmed Zewail receiving Nobel Prize in Chemistry from
King Carl Gustaf XVI at the Nobel Prize Ceremony 1999

Ninth Rajiv Gandhi Science & Technology Lecture, Bangalore

17 October 2002

Light and Life

by

Ahmed H. Zewail

Nobel Laureate
California Institute of Technology
Pasadena, USA

Mrs. Sonia Gandhi and Prof. Ahmed Zewail

Mrs. Sonia Gandhi's Speech

Nobel Laureate Prof. Ahmed Zewail
Prof. C.N.R. Rao
Ladies and Gentleman,

To come to Bangalore lifts the spirit. To come to the Indian Institute of Science lifts the mind. Bangalore's international reputation as a great IT hub is recent. Its status as a major scientific centre is much older and provided the basis for its present eminence. India's future will be influenced in important ways by the work done in this city. I am delighted to be here and am grateful to the Indian Institute of Science for hosting the ninth Rajiv Gandhi Science & Technology Lecture. My special thanks to Prof. CNR Rao for all that he has done to make today's event possible.

These lectures were established for several reasons. Rajivji, like his mother and grandfather, understood fully that science was the key to creating a modern and prosperous nation, a competitive economy and a rational society. The lectures honour his memory, but also remind us of his vision. They give us a rare chance to hear and to meet some of the most remarkable figures in world science.

Prof. Zewail is one of them, a truly outstanding chemist of our time. His revolutionary discoveries have created a new scientific discipline and earned him the Nobel Prize. I was staggered to learn that Prof. Zewail has developed a rapid laser technique that has been described as the world's fastest camera, through which chemical reactions are observed in the femtoseconds in which they actually occur. This is an amazing achievement. Prof. Zewail is living proof of the daring of the human mind and the limitless frontiers of its imagination. He hails from Egypt, a country with which India has had friendly ties from ancient times. The builders of the Great Pyramids – with their awesome mathematical precision – would have been proud to acknowledge you, Prof. Zewail, as a most worthy successor. To the words of welcome spoken earlier, may I add mine. We are honoured to have you with us.

We live in a time when science is making such astonishing advances. One of the most heartening features of post-independence India has been its commitment to joining the front rank of nations in science. We have built a large number of scientific institutions in the last fifty years. We are better able than before to commercialize discoveries made in our laboratories. Some of our mission-oriented programmes have done us proud, of which our space agency is a notable example. We take justifiable pride in having one of the largest populations of scientists in the world.

But we cannot rest on our laurels. The scientific enterprise requires constant care and attention. The rewards which it offers in return are bountiful.

To be a first rate nation, India must also be first rate in science. It must build on its successes but also learn from its shortcomings. Our future economic success, our national security, our very pride in being Indian, are linked to our scientific capacity and our scientific success. We therefore need to promote science in a big way by funding research at the highest level. We must continuously improve the infrastructure of our laboratories and the quality of our universities.

Our success in the software industry reminds us of what is possible. China has been sending scientists and engineers to Bangalore to learn lessons from the Indian software industry. We, too, must be ready to learn from others.

Where will the financial resources that all this requires come from? Rajivji believed in blue sky research as much as in science and technology for development. During his term as Prime Minister, our R&D expenditure in science and technology increased to one per cent of our GDP. It was his ambition to increase this to two percent of our GDP by the middle of the 1990s. We are still far from this goal.

There is a clear need for business and industry to play a larger role and assume greater responsibility. This is slowly beginning to happen but nowhere near as fast as or on the scale that is needed. In the United States and other industrialized countries, a high percentage of funding for basic research comes from the private sector. We, too, must offer incentives to industry and business to invest both in fundamental research and higher education on a substantial scale in the next five years. Our universities should also be given incentives to go out and seek funding from private sector companies for particular research projects. In the medium term, I hope we will be able to evolve a fruitful pattern of partnership between government, the private sector and our universities, designed to give Indian secience greater prominence and visibility. In this way, Rajivji's two percent target could be achieved, even surpassed. In the short term, however, the government must give Indian science all the support that it needs.

Finally, we have a responsibility as individuals and as parents, I am troubled by the younder generation's diminishing interest in science, reflected in the declining science enrolment in our universities. Perhaps as a result of parental pressure, perhaps spontaneously, more and more of our ablest students are turning to subjects which offer high financial rewards. If India is to be a big global player in the knowledge society and knowledge economy of the future, this trend must be reversed.

Prof. Zewail has chosen as the topic of his lecture the theme "Light and Life." I now have great pleasure in requesting Prof. Zewail to deliver his lecture on one of the frontier areas of chemistry and physics.

Light and life*

Ahmed Zewail

IN this country there is a tradition of prime ministers appreciating and supporting science and technology. From Pandit ('teacher') Jawaharlal Nehru to Indira Gandhi and to Rajiv Gandhi all have shown a commitment to scientific research and its critical role in developing the mind, the society, and the nation. Abdul Kalam, a prominent technologist, is the current President. Rajiv Gandhi believed in extending the science base and not to limit it to a privileged few. In one of his speeches he said, scientific research 'must be supported by a very broad base of people who have scientific learning from which we can draw and reach out to the best people available. We have got pillars that reach to great heights, but they remain pillars – we have to turn them into pyramids'. Incidentally, by mentioning the word *pyramids* in his speech in Delhi, he anticipated by 16 years that an Egyptian, who also believes in building pyramids, would be invited to give the lecture honouring his contributions!

Scientific research is the subject of this lecture, but I wish to focus here on one of its pillars – the value of curiosity-driven research and its impact on our life, the life of the 'haves' and 'have-nots'. For this scientific endeavour, I will demonstrate my point from the study of one phenomenon that has occupied the thinking of humans throughout history – it is the phenomenon of light. What is light?

Twelve billion years ago, give or take a few billion, the big bang took place. In this process at the earliest time, light was an integral part of the creation of the universe. In our galaxy the sun has given light for 4.6 billion years; astronomers tell us that in 10 billion years the sun will shrink and become white hot, a white dwarf, and eventually a dark dwarf – the star will be dead and life here will end. For millions of years, light has defined the life of *Homo sapiens*. Through photosynthesis, light has given us food, energy, and the atmosphere. And using light we communicate information, see the big objects (planets and moons) far from us in the vault of the heavens, and see the small microscopic objects (cells and bacteria) our naked eye cannot resolve. Our life becomes invisible without light. From where does light get this transcending power?

People of ancient civilizations believed in light's miraculous power, a mighty power that deserved to be worshipped. The Egyptians had the first single god, the god of the sun-disk Aton (under the pharaoh Akhenaton), and Hindu teachings repeatedly highlight light and en-*light*-enment. A millennium ago, one of the most important scientific advances made in the study of light was that put forward by the Muslim scientist ibn al-Haytham (ca. 965–1038), known in the West as Alhazen and acknowledged to be the greatest scientist of the European Middle Ages. He was the conceptual pioneer of *camera obscura* and his ideas about light and vision were revolutionary: light must travel in a straight path, at a high speed, and light reflects from bodies and refracts in media; our vision is the result of reflecting light to the eyes, and *not* by emitting light from them. Alhazen began with his observations of and experiments with light, then reasoned towards a theory. Alhazen's masterpiece, *Kitab al-Manazir* (*Treatise on Optics*), remained in Western Europe as the primary work on optics for more than half a millennium and up to the time of Kepler and Newton and even later.

It took nearly a millennium until James Clerk Maxwell in 1864 gave the world the first quantitative description of what light is made of – waves of disturbances of elec-

*Based on the Rajiv Gandhi Science Lecture delivered on 17 October 2002 at Bangalore.

Ahmed Zewail is Linus Pauling Professor of Chemistry and Physics, Laboratory for Molecular Sciences, Arthur Amos Noyes Laboratory of Chemical Physics, California Institute of Technology, 1200 East California Boulevard, Pasadena, CA91125, USA.
(e-mail: zewail@caltech.edu)

tric and magnetic fields. These waves move in space and through time. Furthermore, Maxwell's equations predict the correct high speed of light ($c = 300,000$ km/s), which was first estimated in 1675 by the Danish astronomer Olav Roemer and measured in 1849 by the French scientist A.-H.-L. Fizeau. In elucidating that light is an electromagnetic wave, Maxwell unified the important work of Michael Faraday (1791–1867) on electricity and magnetism and of Thomas Young on the wave nature of light (interference; 1801). As a wave, light has a wavelength (λ) and frequency ($v = c/\lambda$) – this is true for all waves of the entire electromagnetic spectrum, from radio waves to X-rays. With Maxwell's breakthrough, scientists of the day thought that the question of the nature of light was answered conclusively, but there was a surprise in store.

At the beginning of the twentieth century, physics witnessed the development of two revolutionary ideas – quantum mechanics (1900) and relativity (1905) – suggesting that in the world of the very small (atoms) and the world of the very large (with very high mass or very high velocity) Newtonian mechanics would not apply, a real blow to centuries of belief.

In 1905, Albert Einstein recognized the implications of quantization for light – it is made of a stream of particles and comes as bundles of energy ($E = hv$) – the light quanta, called by G. N. Lewis as *photons*. The particle description had been advanced by Isaac Newton (1643–1727) and other scientists even earlier, but in Einstein's view the energy (E) and frequency are related by Planck's constant (h). Einstein was successful in using 'this bundling-of-energy' concept to explain the ejection of electrons from metal surfaces, the photoelectric effect, for which he received the Nobel Prize in Physics – *not* for his theory of relativity! With quantization it was possible to explain a variety of phenomena, including the Raman effect, named after the famed Indian physicist C. V. Raman, who in 1928 observed the scattering of monochromatic light as it passes through a transparent substance.

Considering the two descriptions of light by Maxwell and Einstein, we now view light as behaving partly like (electromagnetic) waves and partly like particles – a duality in its nature! Until today we do not fully understand the meaning of this duality, nor do we really understand quantum mechanics, with its uncertainty, as we do the classical mechanics of Newton, with its deterministic laws of motion. But we know how to operate with the dualistic wave-type and particle-type behaviour of light. Remarkably, the same duality was found for all matter at the microscopic level, and now we speak of atoms as particles and as waves: the momentum (p) of a particle is related to its wavelength by the well-known de Broglie relationship ($\lambda = h/p$).

Why are these new concepts important? Besides being brain-teasing, thought-provoking ideas, they provide the springboard for advances in technology. Without quantum mechanics we would not have developed the transistor, the semiconductor industry, and the computer revolution. Neither would we have had the laser, optical communication, and the age of information technology. There would be no global economy to speak of. It is said, notably, that more than half of the US economy is based on quantum mechanics. Without quantum mechanics, we would not be able to tune the radio or communicate with a satellite or position a spaceship – we must know the frequency of the waves used and know how to communicate with them using quantum devices. And we must know the frequency and intensity of this bundle of energy, the photons, to perform eye surgery with lasers.

But there is more. Progress in science is made through paradigm shifts to develop new concepts and new techniques. With optical elements such as lenses and mirrors, light does magic, bringing into focus the world of the very small, the very far, and the ephemeral. The light microscope was developed in the middle of the seventeenth century. In Holland, Antoni van Leeuwenhoek (1632–1723) and in England Robert Hooke (1635–1703) made astounding discoveries, including observation of tiny moving creatures in droplets of water, sperms, and the cellular structure of slices of cork. Hooke coined the word *cell*, and his greatest work, *Micrographia* (1665), defined microscopy as a scientific discipline. As a result of these advances, the history of biology has shifted from an emphasis on classifying living organisms and plants to studying the living cells – the exploitation of light produced cell biology as a new branch of science. Molecular biology and genetics are the most recent frontiers reached with the aid of other developments based on the use of electromagnetic waves, those of X-ray diffraction by DNA and protein crystals and nuclear magnetic resonance of macromolecules. Through scientific experimentation the microscope has magnified the world of the very small – microns in size – so it is visible to our eyes. As a result, human medicine has changed forever.

The telescope was invented before the microscope, in the early part of the seventeenth century; some believe that the first optical assembly of this nature was made in 1550. Hans Lippershey, an eyeglass-maker based in Holland had developed telescopes (1608–09) with a magnifying power of about three times. These and later telescopes were made from a combination of concave and convex lenses and the effect produced was understood from studies of the refraction and reflection of light (e.g. those by Alhazen and later by the Dutch scientist Willebrord Snell). Galileo Galilei (1564–1642) was the first to make use of the telescope to visually approach the very far; first, ships in the distance and then the heavens. Through scientific observations (1610) of Jupiter's moons, which revolve around a planet other than the earth, Galileo refuted the long-held dogma of geocentrism, proving that the stationary earth is *not* at the centre of the universe with the planets and sun revolving around it. The geocentric model – Ptolemaic astronomy at the heart of Aristotelian world view – was to be replaced by the heliocentric model of Copernicus (1473–1543), in which the sun is the fixed centre of the universe and the planets, including the earth, are in circular orbits around the sun; Kepler (1571–1630) refined the Copernican model by showing that astronomical bodies follow elliptical orbits in their motion. Without these concepts and techniques we could not launch a spaceship or a satellite or hope to understand our universe.

Galileo used his telescopic observations along with other empirical data to understand mechanics in general and falling bodies in particular. He provided a new way to test a hypothesis and he refuted Aristotle's theory that heavier bodies fall faster than lighter ones. The 'mechanical philosophy' of interacting particles, or 'corpuscles', elucidated by René Descartes (1596–1650), and the concept of the 'mechanical universe' – synthesized in Newton's magnum opus *Mathematical Principles of Natural Philosophy* (1687) – provided the basis for thinking about motion and the mechanics of macroscopic objects.

In contrast, motions in the microscopic world – *the quantum world* – had never been observed in real time because the human eye responds in the slow sweep of a fraction of a second, while microscopic motions charge along at a faster rate than the eye is capable of. These microscopic motions are ephemeral and ultrashort in duration, and we need a telescope that not only brings their very far world up close for observation, but also freezes them in time so we can take snapshots. We needed what we have termed a *femtoscope*, and as with the ordinary light microscope and the telescope, light is the essential element.

At Caltech, we have been interested in this endeavour of developing ultrafast laser light to construct a femtoscope capable of freezing the motion of atoms, to make a motion-picture film with a frame resolution of a femtosecond. A femtosecond is a millionth of a billionth of a second, i.e. 0.000 000 000 000 001 s. You can see that without the Indian zero and Arabic numerals we would not have been able to express in numerical terms the meaning of a femtosecond! A femtosecond is to one second as a second is to 32 million years. In one second, light travels 300,000 km (186,000 miles), almost the distance from the earth to the moon; in one femtosecond, light travels 300 nm (0.000 000 3 m), the dimension of a bacterium, or a small fraction of the thickness of a human hair. In principle, with femtosecond timing, the atom's motion becomes visible, but how can we advance stop-motion photography to reach the scale of the atom?

In the nineteenth century, the motion of animals was recorded for the first time using light shutters and flashes. In France, Étienne-Jules Marey, a professor at the Collège de France, was working (1894) on a solution to the problem of action photography using *chronophotography*, a regularly timed sequence of images. Marey's idea was to use a single camera and a rotating slotted-disk shutter, with exposures on a single film plate or strip that was similar to modern motion picture photography. Marey applied his chronophotographic apparatus in particular to humans and animals in motion, and to a subject that had puzzled people for many years: the righting of a cat as it falls so that it lands on its feet. How does the cat do it? Does its motion violate Newton's laws of mechanics? Does the cat have some special, magical physiology or a command of some weird new physics or what?

By 'slicing time' and freezing the motion during the fall, in the transition state of the righting, Marey was able to answer the questions. First, the cat rotates the front of its body clockwise and the rear part counterclockwise, a motion that conserves energy and maintains the lack of spin, in accordance with Newton's laws. It then pulls in its legs, reverses the twist, and with a little extension of the legs, it is prepared for final landing. The cat instinctively knows how to move, and high divers, dancers, and some other athletes learn how to move in the absence of torque (the pushing force that gives you momentum in one direction or another), but scientists needed photographic evidence of the individual stopped-action steps to understand the mystery. The answer to the puzzle was that the moving body was not rigid, and Newton's laws prevailed. Marey's work and that of Eadweard Muybridge on the horse have changed the way we think of the behaviour of animals (and humans) in motion.

For the world of atoms in molecules, if the above ideas of stop-motion photography can be carried over in a straightforward manner, then the requirements can be identified for experiments in femtochemistry – the field of studying molecular motions on the femtosecond timescale. The contrast in *length* and *timescales* for the motion of the cat and the atom is awesome. For a definition of 1 cm, a cat speeding at 2 m/s requires a time resolution of 0.005 s. But for a molecular structure in which atomic motions of a few angstroms (an angstrom, Å, is 10^{-8} cm) typically characterize chemical change, a detailed mapping of the motion will require a spatial resolution of less than 1 Å (about 0.1 Å). Therefore, the shutter time, or time resolution, required to observe with high definition atoms in motion with a speed of 1 km/sec (1000 m/s) is 0.1 Å divided by 1000 m/s, which equals 10^{-14} s or 10 fs – a million million times shorter than what was needed for Marey's (or Muybridge's) stop-motion photography.

However, such minute time and distance scales for the atom mean that molecular-scale phenomena should be governed by the principles, or language of quantum mechanics, which are quite different from the familiar laws of Newton's mechanics that were used in the description of the motion of the cat and the horse. In quantum mechanics, the uncertainty principle between position in space and momentum, and similarly between time and energy, led initially to the belief that the femtosecond time resolution would not be useful. Moreover, predictions suggested that localization of atoms in space – wave packets – would not be possible to sustain for a long time, even on the femtosecond scale! Finally, there is a fundamental difference in the analogy between femtoscopy of the atom and the millisecond photography of the cat or horse in that in femtochemistry experiments one probes typically millions to trillions of molecules, and/or repeats the experiment many times to provide a signal strong enough for adequate images. Unlike experiments on one cat or horse, the picture would be blurred.

Conceptually, our work in the late 1970s on coherence phenomena and in the mid-1980s closing in to resolve reaction dynamics in real time provided the foundation for thinking about the issues raised above. It became clear that molecules can be made to vibrate coherently and ensembles of molecules can be made to behave in unison. Experimentally, we needed a whole new apparatus, a whole new 'camera' with unprecedented time resolution. We needed to interface femtosecond lasers and molecular-beam technology, which required not only a new initiative but also a major effort at Caltech.

In 1987, we reached our goal of observing, for the first time, Democritus' atom – theorized by the Greek philosopher some 2500 years ago – in motion, and we could describe it on the femtosecond timescale as a classical object like the cat and the horse. The similarity between atomic motions and planetary classical motions brings about an analogy between the femtoscope and the telescope. In reaching the femtosecond domain of the atom, with a scale of a millionth of a billionth of a second, the time resolution of today compared to that of a century ago, with a scale of a thousandth of a second, is like one day compared to the age of the universe.

Historically, coherence was also not appreciated in the realization of the maser (*m*icrowave *a*mplification by *s*timulated *e*mission of *r*adiation). A pioneer in the development of the maser, Charles Townes, who gave the Rajiv Gandhi lecture in 1997, initially encountered objections to his idea that electronomagnetic waves could be made 'purely' monochromatic, objections based on the uncertainty principle. The claim was that since molecules would spend only about one ten-thousandth of a second in the cavity of a maser, it would be impossible for the frequency of the radiation to be narrowly confined. In the event, coherence of photons in the stimulated emission-feedback process removed this concern, and first the maser and later the laser were developed.

With the femtoscope, the breadth of applications emerging from all over the world spans the very small to very complex molecular assemblies and all phases of matter. An example that demonstrates the unity of concepts from small to large molecular systems came from a paradigmatic study made at Caltech on a sibling of table salt (two atoms) and another at Berkeley on the protein molecule of vision (hundreds of atoms). In both, the primary step involves femtosecond motion of the atoms, and we now understand better the remarkably coherent and highly efficient first step of vision at the atomic level.

An especially exciting frontier for femtoscience is in biology. At Caltech we now have the National Science Foundation's Laboratory for Molecular Sciences (LMS) for interdisciplinary research on very complex systems. Among the recent new studies published are those concerned with the conduction of electrons in the genetic material, the binding of oxygen to models of haemoglobin, molecular recognition of protein by drugs, and the molecular basis for the cytotoxicity of anticancer drugs, and of digestion. We are also developing new techniques to observe the behaviour and architecture of these complex molecules – in space and time – using diffraction images, which give the 3D location of all the atoms, all at once! The impact on biology and medicine is clear.

As for technology developments – femtotechnology – there are exciting new developments in microelectronics (femtomachining), femtodentistry, and femtoimaging of cells and tumours, not to mention possible new developments with intensities reaching that of the sun (in femtoseconds!) and duration going beyond the femtosecond (attosecond), and the interface with nanoscience and technology – marrying scales of time and length. The ability to count optical oscillations of more than 10^{15} cycles per second will lead to the construction of all-optical atomic clocks, which are expected to outperform today's state-of-the-art cesium clocks, with a new precision limit in metrology. There is also the potential for using powers reaching 10^{20} watts/cm^2 to induce nuclear fusion in clusters of atoms through Coulomb explosion. And, the possibility for controlling matter on the femtosecond timescale – one day we may direct chemical reactions into specific or new products.

I now come to the epilogue of this lecture. I have tried through the history of one phenomenon, that of light, to show the power of scientific research that Rajiv Gandhi spoke about. A power that affects life itself; it helps us understand our origin as a species, and aids us in shaping the future. In this context, I am concerned about the recent report in *Nature* of London showing India's fall in its scientific research publication rate – in the past twenty years the number of scientific papers has fallen from about 15,000 to 12,000, while China has increased its output from 1000 to 21,000; South Korea's increase in output over the same period is similarly impressive. It is through science and science education that India can maintain its democracy and continue on the road to prosperity. Decades ago Nehru said the following: 'Who indeed could afford to ignore science today? At every turn we have to seek its aid . . . . The future belongs to science and to those who make friends with science'.

From the story I told today, perhaps several lessons may be drawn. First, in curiosity-driven research we really do not know what we shall discover, but in the process of searching, new concepts and new technologies may be developed, some of which will change our world. Science cannot be 'managed', but instead it requires a nurturing and supportive milieu – if provided, success is certain! Secondly, basic research is the foundation for technological advances; together with input from society they form the real triangle for progress. Cloning is a good example – it began as research in many laboratories, then it transformed into a new technology, and now society must address its ethical, moral, and religious dimensions.

The third point to make is the relevance of science to globalization. Science is international, and success in technology depends on research from the entire world community – the evidence for internationalization is clear in the story presented here, as the contributions made were from all around the globe. Globalization will be more effective and prosperity more widespread and fruitful, if science and technology become basic in the platform of national policy. Finally, science education: a culture of science beginning in primary schools is absolutely essential for the progress of society and for the enlightenment of the mind. It encourages the rational approach to the world, the mentality that seeks to question, to explore, and to participate in team efforts. Moreover, science education is at the core of our peaceful coexistence, as pointed out by C. N. R. Rao in his presentation at the Pontifical Academy.

With proper support and independence, I believe that science (and faith) will continue to provide humanity with light, liberty and learning. But science has to go beyond research and development and must become part of our global education in this modern world. The 'haves' must help and involve the 'have-nots' to alleviate poverty and illiteracy and move toward progress. Scientists are in a position to contribute to this earth-saving cause as they do well in their own disciplines, which promote human progress. No words can describe this feeling than those of Rajiv Gandhi: 'As scientists, you have the power to show us the way. You are not only men and women of science, you are citizens of the human race endowed with unique qualities. You are able to understand the physical world better than others. You have the means to transform it. You owe it to mankind that this special gift is used in the service of peace'.

Indeed when we think of peace we must think of Mahatma Gandhi, who showed the best in the human soul. In Stockholm last December (2001), at the celebration of the centenary of the Nobel prizes, I learned that the Committee for the Peace Prize had intended to give the Nobel Prize for Peace to Mahatma Gandhi in 1948, but he was assassinated and no prize was awarded that year. Had he lived one more year he would have received the Peace Prize. (My advice to those deserving ones who are still waiting is to live long enough!)

Mahatma Gandhi's message in life was tolerance and his words still echo in the world today. On the morning of 13 January 1948, this remarkable Indian leader and world peacemaker commenced his last fast, a life-threatening abstinence to encourage India on the path of peaceful coexistence and cooperation among Hindus, Sikhs and Muslims. Just before he broke his fast, the following Hindu verse was read:

> Lead me from untruth to truth
> From death to immortality
> From darkness to light.

Gandhi's light is as powerful for the spirit as nature's light is for life.

I would like to close by reminding citizens of the world of the noble cause that Rajiv Gandhi wished for humanity – the building of (scientific) pyramids in the service of peace!

Received 25 November 2002; accepted 28 November 2002

Book Index[*]

[*]*Relevant references are given for all pieces presented in this volume.*

Corpus Acknowledgments

The publisher would like to thank the following organizations and publishers of the various journals and books for their assistance and permission to reproduce the selected reprints found in this volume:

Article	Organizer/Publisher
Why It Would Be a Big Mistake for the U.S. to Cut Aid to Egypt. *The Huffington Post*, November 5, 2014	The Huffington Post
Don't Cut Aid to Egypt: The Hopeful Case for Supporting Egyptian President Sisi. *The Los Angeles Times*, November 3, 2014	The Los Angeles Times
The Revolution Egypt Needs. *The New York Times*, October 13, 2013	The New York Times Company
Healing Egypt: Three Steps to Unify a Divided Nation. *The Christian Science Monitor*, July 11, 2013	The Christian Science Publishing Society
Egypt's New Year Resolution. *The New York Times*, January 3, 2013	The New York Times Company
Syria: Is the World Waiting for Genocide? *The Huffington Post*, November 28, 2012	The Huffington Post
Egypt's March Toward Democracy. *The New York Times*, May 20, 2012	The New York Times Company
Pillars of Change in Egypt. *The Los Angeles Times*, December 5, 2011	The Los Angeles Times
As Elections Loom, Egyptians Must Unify. *The New York Times*, October 5, 2011	The New York Times Company
A Compass of Hope for Egypt: The New "City for Science & Technology" Is the Aswan Dam for the 21st Century. *The Huffington Post*, June 22, 2011	The Huffington Post
Fund Egypt's Future to Save the Arab Uprising. *The Financial Times*, April 25, 2011	The Financial Times Ltd

Education System Needs Its Own Revolution to Succeed. *The Times*, February 21, 2011 — Times Newspapers Ltd

We Must Unleash the Power of Egypt's Youth. *The Times*, February 16, 2011 — Times Newspapers Ltd

Egypt's Next Steps. *The New York Times*, February 2, 2011 — The New York Times Company

Obama's Sweet Egyptian Date. *The International Herald Tribune*, September 30, 2009 — The New York Times Company

The West and Islam Need Not Be in Conflict. *The Independent*, October 24, 2006 — Independent News and Media Limited

We Arabs Must Wage a New Form of Jihad. *The Independent*, August 24, 2006 — Independent News and Media Limited

We Must Dream *Replacing the Darkness of Ignorance with the Light of Knowledge. The Cairo Review of Global Affairs*, Volume 6, Page 39 (2012) — The American University in Cairo

How to Jump-Start the Post-Revolutionary Era in Egypt. *New Perspectives Quarterly*, Volume 28 [Spring], Page 39 (2011) — John Wiley & Sons, Inc.

A Scientific Revolution. *The Arab Spring Puts the Middle East in a Position to Become a Scientific Powerhouse, but It Needs Help from the West. New Scientist*, April 23, Page 26 (2011) — Reed Business Information Ltd.

Reflections on Arab Renaissance. *A Call for Education Reform. The Cairo Review of Global Affairs*, Volume 1, Page 36 (2011) — The American University in Cairo

Exploring the Changing Landscape of Arabian, Persian and Turkish Research. Foreword, *Global Research Report: Middle East*, Thomson Reuters, Leeds, Page 1 (2011) — Thomson Reuters

Mediterranean Scientopolitics. *Science*, Volume 321, Page 1417 (2008) — American Association for the Advancement of Science

A New Vision for Science and Technology. *The Only Choice for the Arab World.* Keynote Speech, United Nations Forum on Technology, Employment, and Poverty Alleviation in the Arab Countries, UNESCWA/ILO, Beirut, Page 1 (2002) — United Nations-Economic and Social Commission for Western Asia

How curiosity Begat Curiosity. *The Los Angeles Times*, August 19, 2012 — The Los Angeles Times

The US Needs a New Soft Era. *The Guardian*, July 12, 2010 — Guardian News and Media Limited

Science as a Shaper of Global Diplomacy. *The Los Angeles Times*, June 27, 2010 — The Los Angeles Times

We Need a Science White House. *The Wall Street Journal*, April 17, 2008 — Dow Jones & Company, Inc.

Curiouser and Curiouser. *Managing Discovery Making. Nature*, Volume 468, Page 347 (2010) — Nature Publishing Group

The Soft Power of Science. *The American Interest*, Volume V, Page 117 (2010) — The American Interest

Science in Diplomacy. *Cell*, Volume 141, Page 204 (2010) — Elsevier

Science for the "Haves". *Angewandte Chemie, International Edition*, Volume 52, Page 108 (2013) — Wiley-VCH Verlag GmbH & CO. KGaA, Weinheim

The World in Fifty Years. *Revolutions and Repercussions. The Way We Will Be 50 Years from Today*, Edited by Mike Wallace, Thomas Nelson, Nashville, Page 228 (2008) — Thomas Nelson

The Future of Our World. U. Thant Distinguished Lecture, United Nations University in Tokyo; *Einstein — Peace Now! Visions and Ideas*, Edited by Rainer Braun and David Krieger, Wiley-VCH, Weinheim, Page 109 (2005) — Wiley-VCH Verlag GmbH & CO. KGaA, Weinheim

It is Possible. *One Hundred Reasons to be a Scientist*, 2nd edition, ICTP, Trieste, Page 260 (2005) — ICTP, Trieste

Science in the Developing World. Keynote Speech, General Assembly of Third World Academy of Sciences in New Delhi; *TWAS Newsletter*, Volume 14, Page 23 (2002) — Third World Academy of Sciences

Dialogue of Civilizations. *Making History through a New World Vision* UNESCO Public Address, Paris; *Science and the Search for Meaning: Perspectives from International Scientists*, Edited by Jean Staune, Templeton Foundation Press, Philadelphia, Page 90 (2006) — Templeton Foundation Press

Science for the Have-Nots. *Developed and Developing Nations Can Build Better Relationships. Nature*, Volume 410, Page 741 (2001) — Nature Publishing Group

The New World Dis-Order. *Can Science Aid the Have-Nots? Pontificiae Academiae Scientiarvm Scripta Varia*, Volume 99, Page 450 (2001)	The Pontifical Academy of Sciences
Dire Need for a Middle Eastern Science Spring. *Nature Materials*, Volume 13, Page 318 (2014)	Nature Publishing Group
Science in Arab Renaissance. *Nature Middle East*, doi:10.1038/nmiddleeast.2014.5, January 9 (2014)	Nature Publishing Group
Dreaming the Future. Priestley Award Address, Anaheim; *Chemical and Engineering News*, Volume 89 (13), Page 17 (2011)	The American Chemical Society
Franklin's Vision. Invited Address, American Philosophical Society, Franklin Tercentenary Celebration, Philadelphia; *Proceedings of the American Philosophical Society*, Volume 150, Page 542 (2006)	American Philosophical Society
Time's Mysteries and Miracles. *Consonance with Physical and Life Sciences.* Distinguished Guest Lecture, Bibliotheca Alexandrina; *Discovery to Delivery: BioVision Alexandria 2004*, Edited by Ismail Serageldin and Gabrielle J. Persley, Bibliotheca Alexandrina, Alexandria, Page 9 (2005)	Bibliotheca Alexandrina
Science and Technology in the Twenty-First Century. Public Lecture, Academy of Sciences of Malaysia; *ASM Lecture Series*, Akademi Sains Malaysia, Kuala Lumpur, Page 1 (2002)	Akademi Sains Malaysia
Light and Life. Rajiv Gandhi Lecture, Bangalore; *Light and Life*, Gondals Press, New Delhi, Page 9 (2002); reproduced in *Current Science*, Volume 84, Page 29 (2003)	Current Science Association

While every effort has been made to contact the publishers of all reprinted papers prior to publication, we have not been successful in some cases. Where we could not contact the publishers, we have acknowledged the source of the material. Proper credit will be accorded to these publishers in future editions of this work after permission is granted.